THE NATURALIST IN SCOTLAND

In the same series

THE NATURALIST IN
CENTRAL SOUTHERN ENGLAND
by Derrick Knowlton

THE NATURALIST IN DEVON
AND CORNWALL
by Roger Burrows

THE NATURALIST IN THE ISLE OF MAN
by Larch S. Garrad

THE NATURALIST IN LAKELAND
by Eric Hardy

THE NATURALIST IN MAJORCA
by James D. Parrack

THE NATURALIST IN WALES
by R. M. Lockley

THE REGIONAL NATURALIST

THE NATURALIST IN
Scotland

DERRICK KNOWLTON

DAVID & CHARLES
NEWTON ABBOT LONDON
NORTH POMFRET (VT) VANCOUVER

To my daughter Heather, whose young legs struggled gamely up the Scottish hills

ISBN 0 7153 6627 0

Set in 12 on 13pt Bembo
and printed in Great Britain
by Latimer Trend & Company Ltd Plymouth
for David & Charles (Holdings) Limited
South Devon House Newton Abbot Devon

Published in the United States of America
by David & Charles Inc
North Pomfret Vermont 05053 USA

Published in Canada by Douglas David & Charles Limited
3645 McKechnie Drive West Vancouver BC

Contents

List of Illustrations

PLATES

Photographs not acknowledged above are from the author's collection

IN TEXT

All maps, with the exception of Map 5, are based upon the Ordnance Survey Map with the sanction of the Controller of HM Stationery Office. Crown copyright reserved

CHAPTER ONE

An outline of the country

Scenery–Climate–History of the flora and fauna–Distinguished naturalists of the past

SCENERY

BEYOND THE CHEVIOT HILLS lies the country of Scotland, easily accessible in these days of motorways yet still able to tantalise with its seeming remoteness. To the visitor from England, the back door of Gretna offers a mundane entrance and surely the best approach is via the front door, Carter Bar, where the wild desolate moorland evokes the right response from mind and spirit.

In the popular mind the word 'Scotland' may be synonymous with 'Highland', but that is to ignore the extensive lowlands and plateaux; even the mountainous areas have greatly varied scenery.

Scotland is a large country, three-fifths the size of England. From the Mull of Galloway to Cape Wrath is a distance of 274 miles. The greatest width is 154 miles from Buchan Ness to Applecross, although this reduces quickly farther north so that between Dornoch and Loch Broom the width is only 26 miles. The angle at which Scotland lies to the rest of Britain is often not fully realised; and it is difficult to accept that Stirling and Swansea are on approximately the same line of longitude, but such in fact is the case.

The highest mountain is Ben Nevis, at 4,406ft. There are not many more mountains over 4,000ft; but it is a different story where a height of over 3,000ft is concerned. A famous mountaineer, Sir Hugh T. Munro, decided to count the number of peaks over 3,000ft. It was necessary to define a peak, and for this purpose he interpreted a 3,000ft mountain as one which is separated from other peaks by a dip of at least 500ft. A summit which satisfies this description is popularly known as a Munro and there are no less than 276 of them in Scotland.

The largest loch is Loch Lomond, 27½ square miles in area and 23 miles long. The rivers are relatively short and rapid; the longest are the Tay and the Spey, 118 and 110 miles respectively. Waterfalls abound but vary greatly according to the rainfall from season to season. The 370ft Falls of Glomach in Wester Ross are sometimes claimed to be the highest falls in the British Isles, but Eas a Chual Aluinn near Loch Glencoul in Sutherland is 658ft high, four times higher than Niagara.

The mainland is often divided by geographers into three principal divisions, but for our purpose it is best separated into four: the Southern Uplands, the Central Lowlands, the Grampians and the North-west Highlands. This book includes the inshore islands and covers the mainland with the exception of Caithness, which structurally is part of the Orkneys and is to be included in the volume on those islands.

In general, much of the Southern Uplands consists of rounded grass-covered hills exceeding 2,000ft in a number of places and separated by valleys which have a quiet scenic beauty of their own. More spectacular scenery is found in the Merrick Highlands of the Glen Trool National Forest Park, where the highest peak is 2,764ft. In the eastern part the fertile lowland of the Tweed basin is a pronounced physical feature, balanced on the west by a similar but smaller basin of Old Red Sandstone in Ayrshire, where early potatoes are a notable crop. The north-eastern part of the Uplands has more conspicuous hills in the Lammermuirs and Moorfoots, whilst beyond them lie the Pentlands just south of Edinburgh. These last lie north of the boundary fault and are therefore in the Central Lowlands, an indication that the term 'Lowlands' is far from accurate. As well as the Pentlands in the south, several ranges of volcanic hills are prominent in the northern part of the rift valley. A large proportion of the population lives in the Lowlands, where most of the great industries of the country are sited. North of the Firth of Tay the fertile Carse of Gowrie at the foot of the Sidlaw Hills is famous for its soft fruit, which supplies the jam factories of Dundee. North of the Sidlaws a similar belt of lowland, the Vale of Strathmore, runs north-east through Angus.

The Grampian mountains are deceptive at first glance from a speeding motor-car. They are larger than they look. If the motorist should leave his vehicle and foot his way up the slopes, he would soon find out for himself that they are on a mammoth scale. Due to the general north-east/south-west trend of the country, the central and eastern parts are not nearly so highly dissected and have fewer lochs. 'Dalradian Schist' is an omnibus term which covers a variety of metamorphic rock types, some of which give rise to distinctive scenery. Wherever there is limestone the land surface will be green and fertile; and the belt of Loch Tay Limestone is no exception. This is a hard rock which has been subjected to regional metamorphism; but along the shores of the loch it produces the typically fertile pastures, and the roadside verges in summer are bedecked with flowers. Some of the Perthshire mountains, such as for example the Pap of Schiehallion, have striking cone-shaped summits which result from the even weathering of quartzite. In the northern Grampians the granite of the Cairngorm massif is a rather shapeless elevated plateau, whose immensity is not readily apparent when viewed from the Spey valley. The Cairngorms are surrounded by mountains in the Moine Schists which have a similar scenic form of rounded hills, with ling the dominant vegetation on the slopes. Argyllshire in the west has more rugged scenery, although the granite of the Moor of Rannoch produces a relatively flat and featureless moorland. In western Perthshire is the famous tourist area of the Trossachs, a Gaelic word for 'bristly country', an apt description for the craggy scenery of these Schistose Grits.

North of the Great Glen are the North-west Highlands and virtually all this area consists of superb scenery. It is significant that when W. H. Murray selected twenty-one areas of outstanding natural beauty in a survey commissioned by the National Trust for Scotland, no less than seventeen were in the West Highlands and twelve out of the seventeen were in the North-west Highlands. The southern part of this region continues with scenic forms similar to the western Grampians with even more impressive effect. Glen Affric is without doubt one of the finest glens in the whole of Scotland. An added attraction for the lover of solitude is the remoteness of some districts arising from the almost total absence of roads. Knoydart, fronting the Sound of Sleat, and Ardgour, west of Loch Linnhe, are two such areas. The most westerly part of Scotland is Ardnamurchan, a rocky peninsula approached by a winding lochside road whose craggy wildness is offset in summer by masses of rhododendrons and coastal grasslands rich in flowers. Farther north in Wester Ross the scenery has breathtaking beauty. It is not the height of the mountains, which are lower than the Cairngorms, but rather the shapeliness of their peaks and their individual character which, combined with the narrow glens, create an impression of grandeur usually expected only from mountains of much greater height. Some of the loveliest scenery lies between Loch Torridon and Ullapool, with the wild mountains of Torridon, the ancient pine woods at the foot of Ben Eighe, Loch Maree—so often and justly claimed as one of Scotland's finest lochs—the magnificent cone and corrie of An Teallach and the Falls of Measach in Corrieshalloch Gorge.

If the inland scenery is of outstanding excellence, so also is that of the coast, not of this Torridon area alone but of the whole western coastline from the Strath of Arisaig north to Sandwood Bay in the north-west corner of Sutherland. The latter has been described as the finest bay in Scotland, but my own favourite beach is that of the Traigh Sands at Arisaig, where rising from the shell sand are bare schistose slabs pimpled with barnacles and clothed only with a skirt of wrack and a filmy shawl of orange and grey lichens. Not least among the attractions is the seascape, where in the middle distance the long low plateau of Eigg supplies a perfect foil to the serrated outline of Rhum, and where on a clear day the indistinct humps of Beinn Mhor and Hecla on South Uist emerge nearly 60 miles away above the far horizon. There are many beaches of this nature along the western coast, with the clachans (hamlets) of bright white-painted cottages of the crofting communities.

A few miles beyond Ullapool on Loch Broom, the scenery changes to a type which is unique in Britain. Perhaps it is the nearest approach to a moon landscape. The physical pattern here stems from the geology. The appearance of the Lewisian Gneiss is very distinctive, with smooth boulders and low rounded hills interspersed with innumerable lochs and surrounded by extensive peat-hags. Dotted here and there on this platform are isolated hills of Torridonian Sandstone: Stac Polly, Suilven, Canisp, Cul Mor and Cul Beag. The mainland ends with a varied coastline which includes the mighty sandstone cliffs of Clo Mor, the Smoo cave in limestone at Durness, the deep-water inlet of Loch Eriboll and the sand-dunes of Invernaver.

The eastern side of Scotland is markedly different from the west. The hills lack the bold arresting outlines, the narrow ridges, deep corries and precipitous cliffs, and generally have more gentle contours. There are also more extensive areas of relatively level land. In parts of the north, such as in south-east Sutherland, this consists of rough tussocky terrain, but farther south there is good agricultural land in the Black Isle, the Moray Firth coastlands, the Buchan plateau, Fife and the Lothians.

CLIMATE

Scenery and climate are inseparably linked; the land was moulded into its present shape by the far-reaching effects of climate. But the tense of the last sentence is wrong, for this is a continuous if mainly imperceptible process, and there is no final form. In this short summary we are more concerned with the short-term manifestations: the weather. The average Englishman has a fixed prejudice about Scottish weather; he is convinced that it is cold and wet, anywhere and everywhere north of the Border. The absurdity of such a sweeping generalisation can be seen once it is plainly set down in print. It is a meteorological quirk, for which I have seen no explanation, that often when Scotland is enjoying fine summer weather the rest of Britain is experiencing rain.

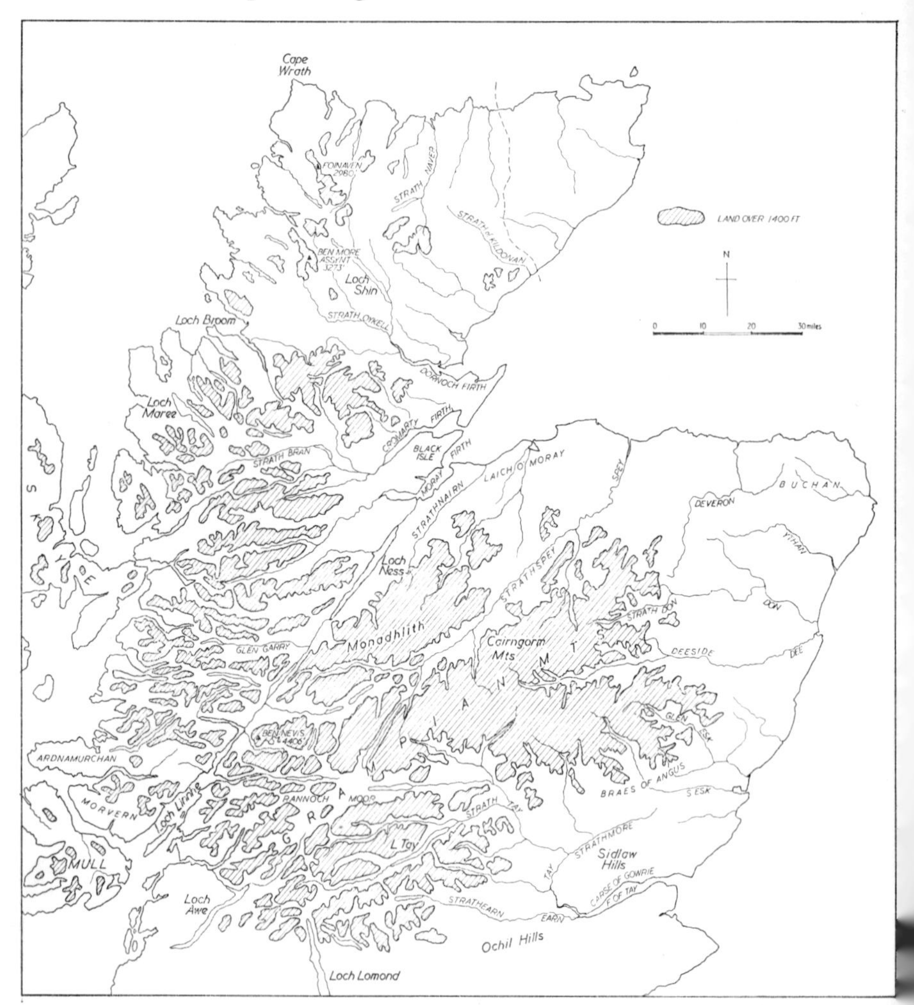

Fig 1 Map showing principal physical features of the Grampians and North-west Highlands

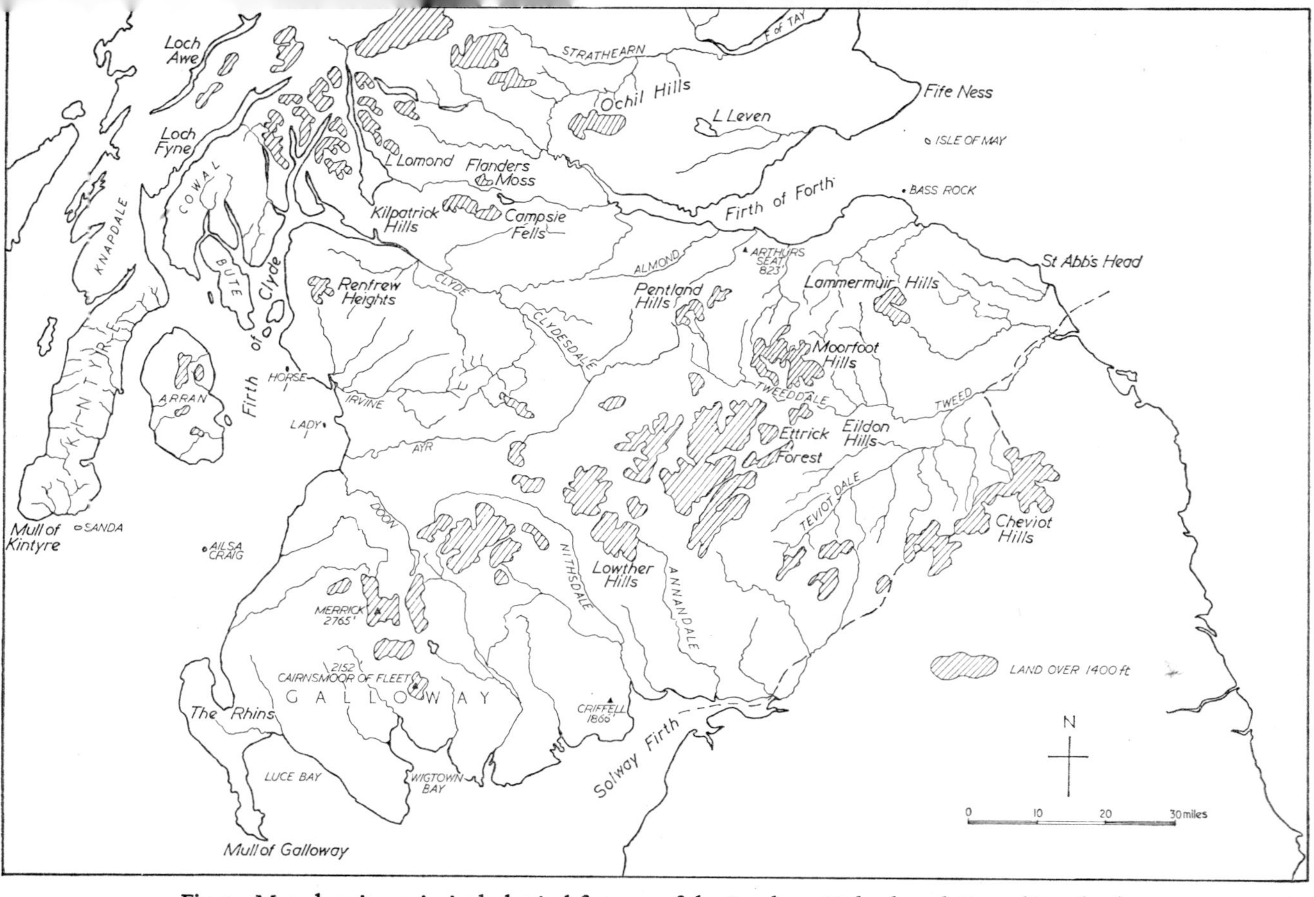

Fig 2 Map showing principal physical features of the Southern Uplands and Central Lowlands

In general, rainfall in Scotland is relatively low, about 45in per annum along a narrow belt of the low-lying western seaboard; but it increases rapidly in the mountains of Argyllshire, western Inverness-shire and Wester Ross, reaching a maximum of nearly 200in in the mountains north of the Great Glen. From here it steadily decreases eastwards, until on the east coast it is less than 30in. The straths and glens naturally have less rain than the high tops, so that areas like the Spey and Dee valleys in the heart of the Grampians have only 30–40in, on a par with the downlands of southern England. During the summer eastern Scotland is warmer as well as drier, so that, for example, although the Tweed basin is on a more northerly latitude than the Merrick Highlands, it has less rain and wind and more sunshine than the latter. The greater amount of sunshine in eastern Scotland is offset to some extent in coastal areas by summer sea-mists, for which the Norse term 'haar' is used, although this phenomenon is much more prevalent in parts of Scotland which lie outside the scope of the present volume. A characteristic of the West Highland climate is that, although the rainfall is heavy, there are frequent periods of sunshine between the showers.

Strong winds occur in western areas, particularly during winter on exposed coasts and mountains: this factor has a direct effect on tree-growth, which is prevented on the small offshore islands and on the mainland above 1,500ft on average in the case of pines and a little higher for birch. As is to be expected, the tree-line is lower in the west due to the prevailing south-west winds; there is a gradation from the 1,000ft contour in the west to about 2,000ft in the Eastern Highlands. Even on lower ground potentially suitable for tree development, windblow has been a significant factor in the life of conifer plantations, as witness the disastrous damage caused at Loch Ard Forest, Argyllshire in April 1967 and again in January 1968.

Because Scotland is on a more northerly latitude, the temperature there is naturally cooler than in the rest of the British Isles, but it is distinctly milder than that of other countries on the same latitude. This is due to the moderating influence of the sea

Page 17 (*above*) The granite massif of the Cairngorms from the north with Rothiemurchus forest in the foreground and the conspicuous cleft of the Lhairig Ghru pass behind; (*below*) the Pap of Schiehallion, Perthshire, composed of Dalradian metamorphic rocks. Note the cone-shaped weathering

Page 18 (*above*) Geological wall at Knockan Cliff showing the eight types of rock present in the cliff face above; (*below*) fossil forest in Victoria Park, Glasgow showing trunks of Carboniferous Sandstone, 230 million years old

and in particular to the effect of the Gulf Stream which flows northwards along the west coast. There is a very narrow temperature range in north-west Scotland, with an average of some 40° F (4·5° C) in winter and 54° F (12° C) in summer. Even in eastern Scotland the average summer temperature does not exceed 60° F (15·5° C). It is here that the greatest extremes of temperature occur. On a warm June day in an idyllic spot in the Spey valley, when it was not easy to imagine arctic conditions, a shopkeeper told me that in the previous winter, outside his shop, there had been a long line of lorries immobilised by the extreme cold.

Frosts are commonplace. The Central Grampians have on average 100 days of frost and on low ground in eastern Scotland there are usually only eight frost-free weeks in summer. Up in the mountains there may be frost at any time. Soil-creep or solifluction is common in the Highlands, chiefly on the higher levels, but in some places as low as 1,000ft. This is the movement of soil particles and rock fragments caused by alternations of freezing and thawing. It produces small-scale physical phenomena such as clitter, that is fragmented rock, and stone polygons in which rock debris is redistributed in angular shapes. In tourist-frequented regions, these features can be seen on Ben Lawers in Perthshire and on Ben Mac Dhui in the Cairngorms. The instability of the soil structure, due to frequent movement, means that it is difficult for plants to obtain a foothold, and in consequence liverworts such as *Gymnomitrium* spp are often the only plants to be seen, although where there is only limited solifluction, plants such as the woolly fringe moss *Rhacomitrium lanuginosus* and the crowberry *Empetrum nigrum* can sometimes be found.

In the Central and Eastern Highlands, snow is frequent; and in the Cairngorms, fresh snow sometimes falls as late as June and as early as September. Snow has an important and beneficial influence on certain types of mountain plant, such as mosses, by protecting them from wind and frost. When the snow melts in early spring, it is amazing how quickly the alpines burst into flower. Snow-bed plant communities occur on a small scale in the Southern Uplands; but as is to be expected, they are much more prevalent farther north, in particular in those areas which have a large gathering ground behind them from which the snow can drift. The north-facing corries of Braeriach in the Cairngorms are a well-known example of semi-permanent snowfields.

HISTORY OF THE FLORA AND FAUNA

The climate of the Ice Age had a direct bearing on present-day plant and animal life in Scotland. Indeed, the history of the flora and fauna cannot be traced farther back than the final glacial phase of the last glaciation. The alternation of relatively warm periods with the glacial ones involved periodic migrations, as arctic species retreated from the increasing warmth or advanced with the onset of colder conditions; those of temperate climate responded in the opposite manner. There can be no doubt that some alpines which occupy the high rock-ledges and summit plateaux were once very much more numerous and more widely distributed. Today they are relict species, clinging with a tenuous grip to the relatively few sites where physical and climatic conditions enable them to do so. Some of course are more successful than others. A few, among which are the tufted saxifrage *Saxifraga caespitosa* and the drooping saxifrage *S. cernua*, are only just managing to survive; although it is good to learn that the former, in addition to its two stations in the Cairngorms and the Nevis range was found not long ago on An Teallach in Wester Ross.

In the warm interglacial periods, plants not only descended on Scotland from the north but also invaded from the mountains of Central Europe, so that existing mountain plants can be classified as arctic, alpine, or in the case of species common to both areas, arctic-alpine. Most Scottish alpines belong to the last group, since in determining the distribution of a plant, climate and soil conditions are more important than altitude alone. With the final disappearance of the ice, a tundra-type vegetation appeared with sedges, grasses and dwarf ericaceous shrubs, together with the typical, arboreal species of dwarf birch *Betula nana* and the various low-growing willows. With continuing improvement of the climate, birch woods were formed in due course; these were succeeded by pine forests, which were probably very open in character, of nothing like the density that we associate with the coniferous plantations of today. Peat deposits indicate that, about 3,000 years from the end of the Ice Age, extensive stands of hazel developed on a scale much greater than that of the present time. In the tree-layer, however, deciduous species were beginning to emerge: especially oak, with perhaps the sessile oak *Quercus sessiliflora* predominant. Two thousand years later came a major, if gradual, change to the wetter Atlantic type of climate which is experienced today, and this was characterised by the presence of alder in abundance and by the formation of many bogs. In the southern half of western Scotland, mixed deciduous woodland was prevalent. A temporary return to a boreal climate, and Neolithic men with their stone axes, combined to make great inroads into the forests. With the development of agriculture, men continued these clearances in order to obtain land both for grazing animals and for crops. The timber was in any case needed for such purposes as iron-smelting, home-building and fuel; oak bark, with its abundant tannin, was required for the treatment of hides. The increasing importance of timber led to plantings, and the area under woodland began to increase again. Coniferous species gained popularity in view of their more rapid growth, the fact that conditions were suitable for them and because with the importation of material for tanning hides and the cessation of wooden shipbuilding, oak woods had become of less economic value.

The establishment of large sheep-farms by the middle of the eighteenth century was a major factor in the story of changing landscape patterns and vanishing wildlife. With the increasing population of Britain came the need for more mutton and wool. The destruction of the clan system, the realisation that sheep could winter satisfactorily out of doors, the more settled times in the Highlands and the desire of the lairds for ready cash, all combined to meet that need. Northern English flock-owners drove their Black-face sheep over the Border to rent vast areas of land in the Southern Uplands. As the lucrative nature of these transactions came to the ears of estate owners farther north, more and more land was turned over to sheep-farms and within fifty years the process had reached Sutherland. The native brown-fleeced sheep, known as Moorit and still seen in small numbers in the Shetlands today, were ousted by the Black-face breed. The effect on the countryside was immense and far-reaching. Many of the remaining woods were clear-felled to make room for yet more sheep. Heavy sheep-grazing on heather moor tends to convert it to grass moor. The burning of heather in the early spring is a regular feature of sheep areas, because it encourages the young growth not only of heather stems but also of grass tussocks. This practice is essential to the farmers, but obviously it destroys some wildlife and it is liable to increase the spread of bracken.

Prior to the coming of the sheep, cattle were reared in quantity in western Scotland; and in the Grampians 'transhumance' was the custom, the cattle being taken up on the hills in the summer months and whole families staying with them, living in shielings. Remains of one of these structures can still be seen on Ben Lawers. The establishment of these temporary nomadic populations with their grazing animals must have had an effect on the natural scene: for example, in the trampling of paths, the arrival of commensals and the greater check on bracken brought by cattle as opposed to sheep. This altitudinal migration each summer ceased when the cattle were replaced by sheep.

The most important event resulting from the arrival of sheep has not yet been mentioned. Up to the mid-eighteenth century, man, as it suited him, had ruthlessly destroyed the lesser creatures over which he had dominion; but now the rewards from sheep-farming were such that landowners cold-bloodedly drove out whole populations from the glens in order to provide wintering ground for the sheep. These enforced migrations, the infamous 'Highland clearances', were bound to have their own effect on natural history, both in the depopulated zones and in the immigration areas on the raised beaches of the west coast. All in all, the development of sheep-rearing led to a general impoverishment of the flora and fauna.

The most recent human activity to affect wildlife is tourism. Coastlands are eroded and made barren, and rare birds are sometimes disturbed in their breeding territories. But these pressures can to a large extent be controlled and Scotland is big enough to be able to accommodate more tourists without irreparable harm being done to the environment. More serious problems, however, are looming up for the future. Vital decisions will have to be made about North Sea oil development. It is devoutly to be hoped that adequate regard will be paid to the country's precious wildlife heritage, even though it must be recognised that economic expansion is sorely needed.

It is difficult to be conclusive and opinions differ, but probably in the main the present fauna of Scotland, like the flora, colonised the country during the last glaciation of the Pleistocene period. As with the flowers, it is the relict species which hold the greatest interest for the naturalist, although—or perhaps because—they form only a small proportion of the total number of species. So, due to the Ice Age, twentieth-century naturalists can listen to the alarm rattle of the ptarmigan on the high tops; watch the northern emerald dragonfly on its beat above the peat-bogs and mountain ringlets fluttering over Grampian slopes; observe the northern dart caterpillar feeding on the linear leaves of crowberry; search by the lochside for the shining brassy elytra of *Pelophila borealis*; or turn over stones to locate other arctic ground-beetles.

In the larger moths, more than twenty species originated from either arctic or alpine areas and the isolating of some creatures in pockets of land or water means that over a long period of time subspecific characters have emerged. The char was a fish of arctic seas which during the glaciations invaded Scottish rivers.

Moving to the most acceptable environment, that is the depths of some of the lochs, it gradually became acclimatised to fresh water and was able to survive when, with the ending of the Ice Age, its way of retreat was cut off. As was to be expected, char became isolated in a number of Scottish lochs, evolving over a period of more than 100,000 years into a number of individual races, some with fairly distinct characters. Another fish genus which was formerly marine but which appears to have undergone a similar adaptation, with members in Loch Lomond and Loch Eck, is the so-called freshwater herring or powan. D. P. Beirne in *Origin and History of the British Fauna* traces the fascinating story of the evolution of hares, from the separation of the mountain hare from the brown hare to the development of a Scottish subspecies in the former and a British subspecies in the latter.

Whatever the origin may be, and it must be remembered that the evolution of subspecies may have causes other than glacial isolation, a number of small mammals have evolved into subspecies. The short-tailed vole on the hills is different from that in the valleys and the lowlands. The fieldmouse found on Bute is a separate race from the mainland animal, and the water-vole of the Highlands is a distinct subspecies, darker in colour than is the type. Butterflies such as the large heath, brown argus, marsh fritillary and chequered skipper have geographical races or subspecies in Scotland. It is, however, in the islands where races, particularly of small mammals, proliferate due to insularity; but that is outside the scope of the present volume.

An interesting feature in the case of a few species is the development of a south-north cline. A well-known example is the distribution of the bridled guillemot. This is a form of the common species, with a white line encircling the eye and extending backwards and downwards towards the nape, giving a distinctly bespectacled appearance. Although detailed consistency does not exist, the overall pattern is quite clear, as indicated by the survey made by H. N. Southern in 1938–9. The percentage of bridled birds at selected places was: Mull of

Galloway 2 per cent; Isle of May 5 per cent; Banffshire coast 6·2 per cent; Handa 10 per cent. It was noted that on the west coast the percentage gradient was not a gradual one, but within the area covered by this volume it was subject to an abrupt increase in the proportion of bridled birds at the Mull of Kintyre. Another example of colour dimorphism is seen in the northern spinach moth. In England the forewings are yellow and the hindwings white; over most of Scotland the colour is darker, with rich brown forewings and smoky hindwings, whilst in north Scotland near-black forms occur.

The amelioration of the climate after the Ice Age, and the development of man into a pastoralist and agriculturalist, had an inevitably deleterious effect on the larger animals. Some of them impeded man's progress and he eliminated them. Others provided food and clothing for his needs but no attempt was made to conserve the stocks, so that they too became extinct. Later, shooting for sport and the collecting mania brought more extinctions. The improvement of the climate after the last glaciation extinguished mammals such as the mammoth, woolly rhinoceros, musk-ox, cave-bear, giant fallow, arctic fox and lemming. The red deer of Pleistocene times was replaced by the smaller form of the present day. The elk, reindeer, wild cattle, bison, brown bear, wolf, beaver and the large sea-bird, the greak auk, survived into historic times. The bison and wild cattle were probably the first to disappear, in the early Iron Age. The bear may have existed until the tenth century, the reindeer and elk a little longer and there were still beavers in Inverness-shire in the sixteenth century. Wolves are cunning animals and as there were still large tracts of forest in which they could take refuge, they survived longer, even with every man's hand against them; by the sixteenth century they had become a menace to the population, and from that time onwards determined efforts were made to exterminate them, even to the extent of burning the sheltering forests. The well-known placename Spittal of Glenshee on the Devil's Elbow road from Blairgowrie to Braemar refers to the refuge provided there from wolves for benighted travellers. The last Perthshire wolf was killed at Killiecrankie in 1680 by Cameron of Lochiel himself, and clearly by this time only a few were surviving in remote fastnesses. Sutherland's last specimen was destroyed by a hunter named Polson in the year 1700, near the entrance to Glen Loth in the eastern part of the county. A stone to commemorate this was erected in 1924 and stands on the bank on the western side of the road at Lothbeg: it is inconspicuous and can easily be missed unless looked for. The exact date when the last Scottish wolf was killed is doubtful, but tradition claims that it died at the hand of a hunter by the name of Macqueen, near the river Findhorn in Morayshire, as late as 1743. This is the more surprising since no wolf had been reported from the neighbouring county of Banffshire since one seen a century earlier, in 1644, at Kirkmichael. Of the smaller mammals the polecat was ruthlessly persecuted and it is very doubtful whether there are any left in Scotland today. The disappearance of the great auk is a tragedy, for it is extinct in the world. It seems incredible that as late as 1840 the ignorance and superstition of men should make them afraid of a harmless sea-bird, but such was the sorry situation and during that year three frightened men of St Kilda killed the last great auk in Scotland, believing it to be a witch. The sea-eagle nested on mainland cliffs until the recent century, probably the last nesting on the mainland being in 1911.

Many species which were not exterminated were greatly reduced in numbers. At the beginning of the present century, a mere handful of pine martens survived in the far north of Scotland, and these were conserved only through the far-sightedness of the estate owner. Birds of prey in particular were subjected to ferocious campaigns during the last century, when it was the custom to pay a guinea for each adult eagle killed; Dr J. Richie records that in seven years, between 1819 and 1826, a total of 355 eagles were killed on just two estates.

But even during these long campaigns of destruction a countermovement was under way, an attempt either to re-establish certain species or to introduce alien ones. Those who made the early introductions were not necessarily motivated by a desire for the welfare of the species for their own sake; but at least it was a welcome step in the right direction. The capercaillie, the largest game-bird in Britain, became extinct in Scotland about 1750. Early attempts at reintroduction in the Dee valley in 1827 were, rather surprisingly, unsuccessful. Ten years later further stock was brought over from Sweden and released on the Taymouth Castle estate in Perthshire. This time came success; and twenty-five years later it was estimated that there were 1,000 birds on the estate, while of course by that time birds had radiated outwards along the glens.

Wild boars were introduced into the Forest of Mar on Deeside about 1830, but they did not long survive. In the eighteenth and nineteenth centuries two introductions of reindeer were made, one in Perthshire and the later one on Deeside, but both were unsuccessful. In 1952 a small number were brought over from Swedish Lapland to the Cairngorms, initially on an experimental basis, and small additions to the herd have been made in subsequent years. Fallow deer were introduced at an unknown date early in the Middle Ages, and the population is known to have been increased subsequently by a number of the black variety brought over from Norway. Japanese sika deer, originally introduced in certain parks, have escaped in several places and have established feral populations. Escaped musk-rats colonised some Scottish rivers in the 1920s, but a determined campaign successfully exterminated them within ten years or so. Mink are now at large, and the present situation will be discussed in Chapter Six.

DISTINGUISHED NATURALISTS OF THE PAST

Even the briefest of visits to Scotland will show the stranger that here is a land whose people are deeply conscious of their heritage. The history of the nation is written in stone in city square and village street, on desolate moor and mountain summit. From the statue in Edinburgh of a mongrel dog perpetuating canine devotion, through the battlefields of Dunbar, Bannockburn and Culloden to the Wallace Monument high on its doleritic outcrop at Stirling, and the outsize statue of the first Duke of Sutherland rising skywards from the summit of Ben Bhraggie at Golspie, the past refuses to lie buried and is constantly invading the present in a direct confrontation from which the casual onlooker cannot escape. It is fitting, therefore, that we should pause for a brief look at some of the distinguished naturalists of the past. For its proportionately small population, Scotland has had a goodly share of them.

It was not until the nineteenth century that any consistent, dedicated study of nature really got under way; and it is noteworthy that often men of this time were not restricted by narrow specialisation but were captivated by a vision of the breadth and oneness of nature. Thus geologists studied wild flowers and botanists recorded incidents of bird life. They can be separated into three broad social groupings. First there were the largely self-educated small tradesmen and gamekeepers, some of them living lives of great poverty but possessed of a tremendous enthusiasm which drove them out to spend long hours in the field after their arduous daily work. Then there were the well-to-do sportsmen and estate owners, leisured men living in a leisured age, with the time and scope to observe nature as they pursued their other interests. Lastly there were the scientific workers from the universities and museums. It is right that the following selective account should be primarily of Scottish men and women; but where an English person has made an outstanding contribution to the study of Scottish wildlife, he is included.

William McGillivray was one of the third group, an early scientific naturalist. He was born in Aberdeen in 1796, but although he received his education there, his childhood home was on the island of Harris in the Outer Hebrides. Incredible as it may seem to the students of today, in common with other impecunious scholars he had to make his way home on foot when each term ended. For him this involved crossing the whole breadth of Scotland several times a year to and from a boat on the west coast, and these long walks must have given him unparalleled opportunities of observing wildlife. After marriage at the age of twenty-four, he spent twenty-one years in Edinburgh and then was appointed Professor of Natural History at Marischal College, Aberdeen. A further momentous event in his life occurred when he met the great American ornithologist John James Audubon. As a result of this meeting he agreed to collaborate by supplying the text to accompany reproductions of Audubon's fine paintings in *Ornithological Biographies*. Books written by McGillivray on his own included *The Natural History of Deeside* and a five-volume work entitled *History of British Birds*, as well as a book each on geology and botany. A bare recital of such a list reveals the author as a general naturalist, but that is not the whole story. He was that rare type of scientist, one who was skilled in both field and laboratory work in balanced proportions. More than that, he was a writer of great sensitivity, and unlike some inhibited modern scientists, he was not afraid to use his imagination to produce wonderfully descriptive prose. The scientific research for which he will always be famous was his original work on the classification of birds. He died in 1852, aged only fifty-six, all in all a man of exceptional rarity, the 'compleat' naturalist.

Colonel H. Drummond was a representative of the second group, and a member of the aristocracy. Born in 1814, a grandson of the Duke of Atholl, he developed in due course an interest in wild flowers and birds. He was a leading figure in the management of the county museum at Perth and an active participant in the affairs of the Perthshire Society of Natural Science, a society still in existence today. When the British Ornithologists' Union was first formed in 1858 he was made its first president, and he is reputed to have been the last man to see a great auk alive—apart, that is, from the man who killed it in Iceland in 1844. Another estate owner was Sir William Jardine, who lived from 1800 to 1874. Like many of his class he was devoted to field sports, but it is as a naturalist that he has achieved distinction. His literary and administrative abilities enabled him to work successfully as editor of *The Naturalists' Library*, which extended to forty volumes.

The small village of Dunipace in Stirlingshire near the Firth of Forth saw the birth in 1844 of another sportsman who was to become probably the best known of Scottish naturalists. John A. Harvie-Brown was a man of great literary industry, and he wrote partly by himself and partly in collaboration with others, a fine regional series of Vertebrate Faunas which extended to nine volumes. The series was never completed, however, and it is ironic that he was not able to carry out his intention of writing a book on his own district, which was perforce left to others to do. His primary interests were mammals and birds. He died in 1916.

John Guille Millais, a distinguished son of the eminent artist Sir John Everett Millais, was born in London in 1865. He too became an artist, and also an animal sculptor and author. He was a Fellow of the Zoological Society and travelled extensively, building up a collection of several thousand birds. He spent several months of each year in Scotland, where he indulged his propensity for field sports. He wrote a number of books on mammals and wildfowl, including *The Wildfowler in Scotland*, published in 1901.

Thomas Edward lived from 1814 to 1886. An Englishman, he spent most of his life in the county of Banffshire. His parents were poor and as soon as he was able to earn a living he took up the trade of shoemaking. Largely a self-taught man, he developed a love of wildlife and as he had great physical energy, he used to roam for many miles over the Banffshire countryside, becoming a practical field naturalist of considerable ability. He compiled a list of the birds of the county, was a competent botanist and discovered twenty species of crustaceans new to the British list.

There are a few people who do not fit into the three social groups mentioned. The Rev Dr David Landsborough was an Ayrshire clergyman and naturalist whose intense interest in the countryside led him to make some seventy additions to the list of Scottish plants and animals. A rare species of seaweed growing in deep water, *Ectocarpus landsburgii*, has been named after him. W. Evans, born in 1851, was a very able field naturalist interested in a wide spectrum of natural history. Ill-health compelled him to retire from a professional post at the early age of forty-two, and he devoted his long years of retirement, until his death at seventy-one, to the serious study of nature. One of his notable achievements was the collecting of a large number of spiders which he supplied to English arachnologists, thereby spreading a greater knowledge of the content and distribution of the Scottish spider fauna. Not only did he add some species to the British list but he discovered some completely new to science, two of which, *Evansia merens* and *Caledonia evansi*, have been named in his honour.

The complexity of Scottish geology has taxed the intellectual resources of the greatest British geologists. James Hutton was born in 1726 into a Berwickshire farming family. At Edinburgh University he studied chemistry and medicine but in the next few years vacillated between medicine, farming, travelling on the continent and studying chemistry. Eventually he returned to the country of his birth and began studying the rocks in Glen Tilt and other places. He saw that an understanding of the nature of geology could be reached by observing the natural weathering processes in action every day. This interpretation was quite at variance with current views and even his epoch-making two-volume work, *The Theory of the Earth*, published in 1795, made little impact. It is now recognised that Hutton occupies an important place in the history of geological science.

Sir Charles Lyell was another great Scottish geologist; following Hutton, he consolidated and developed his teaching. He was born in Angus, of wealthy parents, in 1797. He was a collector of butterflies and moths as a schoolboy living temporarily in the New Forest, but as a teenager browsing in his father's library his eye lighted on a book on geology. Having read it his interest was aroused and this subject soon became his main passion. When the family returned to Scotland he explored the rocks of a number of counties and subsequently wrote a famous book, *The Principles of Geology*, based on the views of his predecessor, Hutton. Knighted by Queen Victoria at Balmoral, he died in 1875 at the age of seventy-eight.

Two men, one a Scot, the other an Englishman, achieved fame in the study of Scottish fossils. Hugh Miller was born in 1802 near the shores of Cromarty Firth. His birthplace is now in the care of the National Trust for Scotland and is open to the public. He began earning his living in a quarry and although even as a child he had been interested in coloured stones, his serious study of geology began after he had casually tapped open a nodule to find that it contained an ammonite. But Miller's name will be forever associated with the fossil fishes of the Old Red Sandstone; he wrote several books, a well-known one being entitled simply *Old Red Sandstone*. Due to his initiative, palaeontologists working in other areas of this rock type were inspired to search for fish fragments, as a result of which there was a rapid extension of knowledge in vertebrate palaeontology. A fossil sea-scorpion, *Hughmilleria*, was named after him. He died aged fifty-four.

The Englishman, Charles Lapworth, came from Berkshire. As a young man in 1864 he obtained a teaching post in Galashiels and whilst there developed the hobby of fossil-hunting. He found an abundance of graptolites, those exceedingly odd and now extinct creatures which look more like serrated pieces of metal than the homes of living creatures. He soon found that there were many different species contained in the various rock strata, and by building up a large collection he acquired a detailed knowledge of the group. But his notable contribution to geological science was his recognition that the apparently erratic disposition of the graptolite species indicated that the Ordovician rocks of the Southern Uplands had been subjected to enormous pressures, with many resultant folds. Once this realisation had dawned on him, he was able to ascertain and describe the structure of the Southern Uplands. Largely as a result of this research he was appointed to the post of Professor of Geology at a college in the Midlands. His name is preserved in the scientific name of a Silurian brittle-star, *Lapworthura miltoni*, and a Cambrian annelid worm, *Lapworthella*.

The brothers Sir Archibald and Professor James Geikie were both distinguished Scottish geologists. The former (1835–1924) was the better known, becoming Director-General of the Geological Survey of Great Britain, and being appointed President of the Royal Society in 1908. He wrote a number of books on geology, but the one which is of general interest is *The Scenery of Scotland*, published in 1887. In his work for the Geological Survey in Scotland he had several gifted geologists working under him. Two such were B. N. Peach and J. Horne, whose work in unravelling the complex structure of north-west Scotland is commemorated by a memorial stone at Loch Assynt.

Sir Robert Sibbald was one of Scotland's early naturalists. His book on the natural history of Scotland, *Scotia Illustrata sivi Prodromus Historiae Naturalis*, was published in 1684. Dealing mainly with botany, it earned him the title of father of Scottish botany. Sibbald helped to establish Edinburgh's famous Royal Botanic Garden and his name is perpetuated in that mountain plant of the rose family *Sibbaldia procumbens*. He is considered to have been instrumental in stimulating a greatly increased knowledge of Scotland's flora and fauna.

For any extensive knowledge of Scotland's special botanical treasures, the mountain flora, we have to wait for almost another hundred years; in 1778 the Rev John Lightfoot published his *Flora Scotica*. Sibbald's list of 500 plants was enlarged by Lightfoot to 1,250. He was probably the first person to publicise the rich flora growing on the calcareous schists of Perthshire, although this was already known to local botanists. When the English zoologist Thomas Pennant began planning a grand tour of Scotland, he invited Lightfoot to be his travelling companion. In 1772 they travelled the length and breadth of the country, including most of the Inner Hebrides, taking a good part of a year over the excursion. What more satisfying way could there be of spending a summer in Britain?

If it was Lightfoot who largely discovered the mountain flora, it was George Don, a nurseryman from Forfar, who a few years later made extensive studies of mountain plants throughout the Grampians: from Knoydart in the west to the Cairngorms in the east, but especially in the mountains of his own county. Of necessity he had a commercial interest in collecting plants, but it is clear that he was interested in botany for its own sake. He acquired a detailed and comprehensive knowledge of the arctic-alpine relict flowers and in the process discovered many new plants for Britain. Some of these were new to science, as for example the alpine fox-tail *Alopecurus alpinus* on Lochnagar and the sedge *Carex rariflora* on the mountains of Clova. Ten years before his death in 1814 he published the *Herbarium Britannicum*.

Throughout the nineteenth century the study of plants rather than of vegetation occupied the attention of the Scottish universities, which established for their students a tradition of field excursions into the mountains. An incident which happened on one such botanical expedition is worth mentioning, not for itself but for its wider significance in that it is an early illustration of the problems of access to the countryside and of the divergent views on the rights of ownership. In 1847 J. H. Balfour, Professor of Botany at Edinburgh University, was leading a party of students in plant exploration of Glen Tilt in Perthshire, when they encountered the landowner with his servants who attempted to stop trespass. The resulting fracas ended in a lawsuit, and the unfortunate episode highlights a conflict of interests, the resolving of which to this day still largely depends upon mutual goodwill, respect and tolerance.

With the beginning of the twentieth century two further aspects of Scottish plant life were investigated. Some botanists and palaeobotanists began studying the history of the British flora. An Englishman who became deeply interested in the origin of Scottish mountain plant life was A. J. Wilmott, a deputy keeper in the Department of Botany in the British

Page 35 (*above*) Loch Sunart beach, Argyll with knotted wrack in foreground and channel-wrack growing on barnacle-encrusted igneous rock behind; (*below*) eider duck with ducklings at Ythan estuary, Aberdeenshire

Page 36 (*above*) Sand-dunes of St Cyrus reserve, Kincardineshire; (*below*) an uncommon flower of shingle and boulder-strewn beaches, the oyster-plant

Museum. He lived from 1888 to 1950, and on the mainland studied plants in the Highlands north of the Great Glen, a region which he found of especial fascination. The study of Scottish plant ecology was begun just before the turn of the century by a botanist from Perthshire, Robert Smith, who published papers on his pioneer studies in certain areas. Born in 1874 he had a tragically early death at the age of twenty-six. His brother W. G. Smith, another plant ecologist, continued his work and produced further studies on Scottish vegetation, based partly on his own researches and partly on his younger brother's unpublished notes.

It can be taken for granted that even before the dawn of recorded history men and birds, in Scotland as elsewhere, were involved in complex relationships of varying intensity. We can well believe that Neolithic men guffawed raucously at the comic appearance and antics of the puffin, shrank away in superstitious dread from the mysterious garefowl, resented the insolent daring and low cunning of the Corvids, waged relentless war against their competitors the birds of prey and enjoyed the thrill of the chase as they hunted a variety of wildfowl for food. These primitive emotions remained for thousands of years and still exist, but at an unknown date a few people began to make observations of bird life. If James Fisher's opinion is correct, a Saxon poet in the seventh century was writing vividly about birds on the Bass Rock. Martin Martin was a traveller who, in his extensive journeys near the end of the seventeenth century, noted the bird life around him; parish records in the Old Statistical Account almost at the end of the eighteenth century indicate that a number of the ministers had a considerable knowledge of birds.

Many an unknown shepherd in the heyday of the great sheep-farms and many an unremembered gamekeeper on the grouse moors and deer forests built up a large stock of practical ornithological knowledge. Then, too, the all-round naturalists referred to earlier were birdwatchers to a man. James Grahame was a poet and birdlover. Born in Glasgow in 1765, the son of a lawyer, he became a clergyman who throughout his life observed birds with meticulous care and wrote verses about them. In 1806 he published *The Birds of Scotland* and died in 1811.

The nineteenth century saw the rise of the sportsman. Many of these sportsmen from their encounters on the hill could not fail to become knowledgeable about birds. Some published their experiences. One such was Charles St John, an Englishman, who wrote *Natural History and Sport in Moray*, *A Tour in Sutherlandshire* and *Wild Sports and Natural History of the Highlands*. Born in Sussex in 1809 he was a fish out of water in his first post as a clerk in the Treasury. When twenty-four years old he resigned, and with the assistance of wealthy relatives he was able to settle in Scotland and live the life of a country laird, a way of living much more to his taste. His was a character of contradictions. There is no doubt that he was an accurate observer of birds; equally there is no doubt that in shooting the objects of his interest he was merely living in the manner of his times. He is often criticised, but in fairness to him and those like him it should be noted that the idea of conservation barely existed at that time. What I think most people find difficult to accept is the inconsistency, to say the least, of a man who combined sentimentality with ruthlessness and who condemned the shooting of ospreys whilst continuing to shoot them himself. Due to this persecution, in which he played a conspicuous part, the osprey became extinct as a breeding bird in Scotland for almost fifty years.

Dr William Eagle Clark, who lived from 1853 to 1938, was an Englishman who obtained the post of Natural History Keeper at the Royal Scottish Museum, Edinburgh. He was one of the pioneers of bird-migration studies. He persuaded lighthouse keepers to keep records of the birds they saw and he laid solid foundations for the later development of bird observatories. In 1912 he wrote *Studies in Bird Migration* and his name will always be associated with Fair Isle. Evelyn V. Baxter and Leonora J. Rintoul were lifelong companions so that it seems fitting to treat them as one here. For many years they worked together in a notable ornithological partnership, until Leonora Rintoul died in 1953. Evelyn Baxter survived her friend by six years. Inspired and assisted by Dr Eagle Clark they carried out migration studies on the Isle of May from 1907 onwards and together wrote a number of papers and books. These included *A Vertebrate Fauna of Forth* in 1935—a work which Harvie-Brown had envisaged but had not been able to carry out—and a two-volume work, *The Birds of Scotland*, published in 1953. Scotland continues to be served by a number of able birdwatchers and experts in other fields of natural history.

CHAPTER TWO

Geology

Stratigraphy of the regions—Effects of glaciation—Geology and wildlife

STRATIGRAPHY OF THE REGIONS

IT IS APPROPRIATE that the country which provided one of the last refuges for Iron Age tribes should possess the oldest rocks in Britain. Until recently it was thought that the Lewisian Gneiss of the far north-west was the oldest formation in the world, but it is now known that there are parts of the Lawrentian shield in Canada which are some 2,000 million years older. Nevertheless, much of the land mass of Scotland is still extremely old even in geological concepts of time. It consists in the main of ancient sedimentary and metamorphic rocks, of Palaeozoic and Pre-Cambrian age, through whose surface in some parts of the country Tertiary volcanoes have erupted, leaving their distinctive features of dykes, sills and lava flows. Beds newer than the Carboniferous period are poorly represented; but New Red Sandstone is seen in a few places, and Jurassic clays and limestones in parts of the north-eastern seaboard and the west coast, where there are also small areas of Cretaceous strata.

The area of the Southern Uplands comprises a number of different systems. Silurian sediments cover the largest part from Wigtownshire eastwards to the Tweed basin. North of the Silurian beds is a narrow parallel belt of the older Ordovician series. Both of these formations trend north-east across the region from the Rhinns of Galloway. This north-east alignment originated from the Caledonian earth movements some 300 million years ago which precipitated major faults along this compass-bearing. It is these hard, highly folded sedimentary rocks which produced the massive, rounded grass-covered hills which are so characteristic of the region. Both systems contain a variety of primitive fossils, in particular the graptolites whose remains appear today as rock graffiti, which indeed is the meaning of the name in the Greek from which it is derived. In Galloway, several granite bosses have intruded into the surrounding deposits and have given rise to a wilder, more rugged type of country with heather replacing grass as the dominant vegetation: the largest such mass is in the Merrick Highlands.

New Red Sandstone is found in several relatively small areas in southern Scotland. The adjective 'red' is misleading in this locality at least, for the rocks here are an attractive shade of pink. This can be well seen on the west bank of the river Nith, where in the hamlet of the Sweetheart Abbey men have taken the pink sand and built an abbey: this, though now little more than a romantic ruin, perpetuates to the imaginative eye the sand-grains of the Permian desert.

The use of the term 'New Red' is a reminder that, 100 million years earlier, the Old Red Sandstone was being laid down in shallow lakes in conditions quite different from those of the later deposit. One such basin lies in the eastern part of the Southern Uplands, in the valley of the Tweed, and forms the foundation for a prosperous area of mixed farming. Near Melrose there is a line of volcanic hills, the shapely peaks of the Eildons, whose rocks contain the uncommon blue mineral riebeckite. The lowest part of the Tweed valley, the Merse, has sandstones of Carboniferous age. There are other scattered Carboniferous beds in the region: for example, the coal-bearing measures at Sanquhar; the fossiliferous limestone on the Nith estuary; and the area of former volcanic activity in the Calciferous Sandstone in the vicinity of Dunbar.

Good use has been made in the past of the rocks and minerals of the region; and recently there has been a renewal of interest in the economic possibilities of re-exploiting the mineral wealth described by a nineteenth-century writer in a book with the quaint title of *God's Treasure-house in Scotland.* The area around Wanlockhead in the Lowther Hills has lead and zinc mines which were in operation for many years and had the largest output in Scotland until production ceased earlier this century. They were reopened for a short period in 1959, and consideration is at present being given to reopening some of these mines for other minerals, consequent upon a recent survey. Gold from here was used in the making of the Scottish crown. Greywackes and grits of the Silurian and Ordovician series have been used as building-stones in the houses of the central Southern Uplands; and one does not need to be a geologist to recognise Red Sandstone country, for the colourful blocks of walling are readily seen. The granite quarry at Dalbeattie which is still active, has supplied stone to many buildings and lighthouses throughout Britain; but perhaps its most notable contribution has been in the construction of the Thames embankment. Another quarry at Creetown produces white granite which is used for non-skid road-surfacing.

North of the Southern Uplands is the belt of country which is always known as the Central Rift Valley. The use of this term is sometimes criticised, but one can see how it originated, for the land in question lies between the two highland massifs and is a displacement block on a monumental scale. Yet it is far from being the geographical centre of the country, and it is certainly not a valley, for it includes several ranges of hills. Two faults—in the north the Highland Boundary Fault from Helensburgh to Stonehaven, and in the south the Southern Uplands Fault from Girvan to Dunbar—have lowered by several thousand feet the vast tract of land between them. The effect of these Caledonian earth movements has been that of ensuring the preservation of younger strata such as the Coal Measures by protecting them from erosion. In the south-west, the Ayr basin of Old Red Sandstone is backed by the rim of the Southern Uplands. More deposits of this age occur farther east in Lanarkshire and Midlothian and, due to the synclinal character of the Rift Valley, they outcrop again in the north-east, over a large area, from just north of Glasgow eastwards through the Carse of Gowrie to the Kincardineshire coast. Fossil fish and arachnids are widespread if not plentiful and plant remains have been found in some places. Above this formation lies the Carboniferous series containing the great Scottish coalfields from Ayrshire to Fife.

Arising out of the Central Lowlands is a number of volcanic hills of Old Red Sandstone or Carboniferous age. They are the Renfrew Heights, the Kilpatrick Hills, the Campsie Fells, the Ochil and Sidlaw Hills in the northern part and the Pentlands south of Edinburgh. The capital city itself has much of geological interest (see Appendix, page 201). The southern edge of the Ochils is a fault-line, south of which are younger Carboniferous rocks at a lower level. Several earthquakes have occurred along this line of fracture in recent historical times. Two conspicuous features just south of the Ochils are the quartz-dolerite sills on which are sited Stirling Castle and the Wallace Monument.

North of the Highland Boundary Fault the Grampian mountains largely consist of metamorphosed sediments: the Dalradian series over most of the Central Highland region, with the Moine schists beginning in northern Perthshire and covering a vast area of the North-west Highlands. These two series are separated by a boundary slide, are exceedingly complex and have provided many problems for geologists. Their position in the geological column, for one thing, was for long undecided although there has been general agreement that they are very ancient rocks. It is now considered that the Moine rocks are metamorphosed Torridonian Sandstone and that the Dalradian are possibly altered Lower Palaeozoic. The latter consists of schists, grits, slates and limestones of sedimentary origin, intensely compressed into many, often recumbent, folds and subsequently subjected to violent thrust movements. Some of the mica-schists are highly garnetiferous; most of the garnets are minute, but sizable specimens can be found in a few places. The rocks vary in the degree of metamorphism, and certain minerals which are present in successive stages can be used as zonal indicators demonstrating the intensity of the transformation.

Intruded into the metamorphic schists are several enormous granite blocks around the head of Loch Etive, the Moor of Rannoch, Ben Nevis and the Cairngorms; also Lochnagar and a number of other localities in Aberdeenshire. The Cairngorm massif, totalling 160 square miles, gives an indication of the scale of these intrusions. Certain semi-precious stones, notably beryls and cairngorms, occur in veins in the Cairngorm granite and on the mountain plateau. The Highland granites vary considerably in composition and the pulpit of Crathie church on Deeside is constructed of no less than eighteen varieties of Scottish granite. As well as the acid granite there are basic rocks—chiefly gabbro—in several parts of Aberdeenshire, particularly at Huntly and Insch.

The western Grampians have, in addition to the Dalradian schists and granitic intrusions, deposits of Mesozoic beds. Although these extend now only over small areas of Morven, Ardnamurchan and Applecross, some geologists consider that Cretaceous strata may well have covered at one time a considerable part of the country. The western coastline also has a variety of volcanic rocks of Tertiary age. At the same time as the Eocene sea was flowing over the chalk hills of southern England, violent volcanic activity was bursting out on the Scottish mainland at Ardnamurchan as well as in the Inner Hebrides. There was a common factor at work in these widely separated areas: deep underground the pressures of the Alpine earth movements were active, in England folding the chalk downs into anticlines and in Scotland splitting open the rocks into fissures from which poured layer after layer of basaltic lava now covering 51 square miles in the districts of Morven and Ardnamurchan. A tremendous miscellany of rock types can be seen in the Grampians. Some are attractive in appearance, like the Glen Tilt marble which consists of green serpentine set in a matrix of white calcite.

The Great Glen forms the dividing line between the Grampians and the North-west Highlands. This conspicuous physical feature is the result of a horizontal tearing apart of the earth's crust, in contrast to the Highland Boundary Fault which is a vertical dislocation. The crustal instability indicated by these faults continued into historic times and many earth shocks have been experienced. The strongest disturbance took place in 1816, near Inverness on the Great Glen Fault, when houses were damaged and the shock was felt as far away as the Southern Uplands. Fissures appeared in 1901 on the Caledonian Canal towpath and the latent instability at the site of all the faults is always there.

The Dalradian schists are now left behind and the Moine schists dominate the scene. The main rock types are mica-schists —layered rocks with a silvery sheen due to the abundance of muscovite mica—and granulites which are hard flaggy rocks; but there is a number of other types. The Moine series extends north-eastwards from Ardnamurchan in the south-west to the north coast of Sutherland, interrupted only by a number of igneous intrusions such as the granite of Strontian and the syenite of Ben Loyal. East of the Moine belt there is a large area of Old Red Sandstone around the Cromarty and Dornoch firths; and farther north there is a small outcrop of Jurassic strata at Brora, where there is a coalfield. Not owned by the National Coal Board but run by a local co-operative, it is the only one in Britain in the Jurassic formation and it produces a coal used principally in industry. A few miles north of Brora is the Strath of Kildonan, the scene of a miniature gold-rush in 1896 when placer deposits were found to the value of £12,000.

Along the western coastline of Ross-shire and Sutherland is an area of more Pre-Cambrian rocks: this is the Lewisian Gneiss, between 1,200 and 2,000 million years old. A small part of the Gneiss was sedimentary in origin, but most of the rocks were originally igneous ones which have been highly metamorphosed. These gneissose rocks are the remains of an ancient mountain range which has been worn away by erosion. On top of these hills and perhaps some 600 million years later, Torridonian Sandstones were laid down. The Gneiss therefore became a buried landscape. These sandstones were formed in desert conditions and in places were once 20,000ft thick. This type of rock in general is vulnerable to erosion. Some Torridonian Sandstone, however, is very hard and some of the higher parts were capped with hard quartzite of Cambrian date; the effect of this is to produce a strange landscape of isolated peaks. Part of this Cambrian strata is named Pipe Rock, because it contains the fossil casts of a worm *Scolithus linearis*. Another Cambrian deposit was a yellow, shaly limestone known as the Fucoid bed, because certain markings within the rock bear a slight resemblance to seaweed. These markings are now known to be flattened worm-casts. Then there is another fossil-bearing band of sandstone described as Serpulite Grit and the Cambrian sequence is completed with layers of magnesian limestone.

At an unknown date following the Cambrian era certain intrusions of igneous rock took place in the district. One such intrusion occupies an area of 10 square miles near Loch Borolan and the rock, which is of an uncommon type, has been named Borolanite. Tremendous earth movements occurred 200 million years after the Cambrian era, making the rock structure extremely complicated, particularly in the Ben More Assynt area. The strata have been piled up in incredible chaos, crumpled, folded on top of each other, sometimes turned upside down; and tremendous masses of older rocks have been lifted and thrust forward up to ten miles, in a westerly direction, on top of younger rocks. There are several of these thrust planes and the largest is known as the Moine Thrust.

EFFECTS OF GLACIATION

About a million years ago the Ice Age began, and its effects can be seen in a number of different ways throughout the length and breadth of Scotland today. It is true that the main outline of the landscape was by then complete; nevertheless, the glaciations had a tremendous influence. There is evidence of four glacial periods and it is probable that there were glaciers up to 10,000–12,000 years ago. Nearly all the mountainous areas were covered with ice-caps, although it is considered that one or two summits —for example, An Teallach in Wester Ross—may have been free of ice. From these ice-caps glaciers flowed to all points of the compass. In addition, the presence of Norwegian rock fragments of a type not found in Britain show that ice-sheets made their way across the North Sea from Scandinavia.

The effect of the glaciers was profound. They altered the profiles of the valleys to the characteristic U-shape as they deepened them, leaving tributary burns in hanging valleys to find their way by means of waterfalls down to the parent river in the glen below. Corries were gouged out of the hills and virtually every rock and boulder were abraded by ice. Several different kinds of deposits were carried down by the glaciers. Moraines of various types are widespread: they are accumulations of rock debris transported by glacier movement and are often found in corries where they have caused small lochs to form. Lateral moraines occur along the sides of glens and terminal moraines at the glacier's end. Boulder clays consist of fine material containing stones brought down by ice to the lowlands and these deposits cover large areas of Scotland today. Glacial sands and gravels were deposited in many places. Districts such as Galloway have numerous drumlins (rounded hills of boulder clay) and eskers, which are narrow ridges of glacial gravel. The power of the ice is such that even large rock masses, termed erratics, are transported miles to another area with a different geological formation where, maybe perched on a hillside, their alien nature identifies them for what they are. The underlying rock structure may modify the work of the ice to some extent but, in general, glaciers possess seemingly illimitable power to score with striae the hardest rock, to move hillocks if not mountains, to grind to powder and to destroy. When the ice-sheets finally melted, the release of their weight caused the land level to rise and raised beaches resulted in a number of places round the coastline.

GEOLOGY AND WILDLIFE

The effect of geology on wildlife is often indirect, linked with other causative factors, as remote and hidden as the underground strata themselves. But that it has a profound influence is never for a moment in doubt. The calcium content of the narrow belt of Durness Limestone enables rare lime-loving alpine flowers to establish themselves. The hummocks of moraine and the rocky banks above the roadside in the eastern Grampians provide ideal conditions for gorse and broom. The relative susceptibility to erosion of Torridonian Sandstone coastal cliffs, in contrast to the harder Lewisian Gneiss, means that the sea-bird colonies in the north-west are located on the ledges of the former. On the other hand, but for much less obvious reasons, greenshank find that gneiss provides the preferred habitat. A certain glacial hillock in southern Scotland, well known to ornithologists, is a favourite feeding ground for several species of geese, probably in the first instance because of the richness of the grass but maybe also because the elevated position assists the wary eyes of the geese. The igneous dykes, whose alignment across flowing water produces waterfalls, indirectly supply dippers with their ideal nesting sites in moss under such falls. In similar fashion, the weathering of rock structure which causes overhanging cliffs, creates the right conditions of inaccessibility for nesting birds of prey.

A notable result of the Ice Age was the leaving of glacial relict species as mentioned in Chapter One. It is interesting to note that if there is an unknown creature in Loch Ness, it may well be because of differential movements between land and sea which have blocked the estuary with a land barrier.

There is one other aspect of the relationship between geology and nature study, and that is the action of the rocks in preserving for examination in the open-air 'museums' of quarries, railway cuttings and cliff faces, the flora and fauna of bygone aeons—from fossil forests to fishes of the Devonian seas. There is indeed enough evidence to demonstrate that geology is the foundation of natural science.

CHAPTER THREE

The coastline

Firths and sea-lochs—Rocky shores—Cliffs

CONTEMPLATING THE INFINITE variety and the spectacular beauty of the Scottish coastline, one is at a loss to describe adequately the superlative nature of the scene. Beauty such as this cannot be imprisoned in a word; it must be seen. If one aspect stands out more than another, it must surely be the diversity of loch and shore and cliff. The sea-lochs, for example, are by no means carbon copies of each other. The west-coast inlets not only differ in general character from those on the east coast but are often dissimilar to each other. There are, for instance, the Loch of Heaven and the Loch of Hell: what greater extreme is possible! There are contrasting sands too: muddy sands in parts of the Solway; the golden miles of east-coast resorts; the silica sands of Morar; the shell-sand beaches of Sutherland; the pink sands of Gruinard Bay and the red of Kintyre sandstone or of Buddons Ness garnet; and in one or two localities, the bizarre fragments of so-called coral sand.

One aspect which must appeal to the nature-lover is the relatively unspoilt and uncommercialised nature of much of the coast. For those who appreciate the solitude so well expressed in Byron's words: 'There is a rapture on the lonely shore, there is society where none intrudes, by the deep sea, and music in its roar', there are many miles of lonely shore where, even at the height of the holiday season, one can wander at will and watch the marine life unobserved by the curious eyes of one's fellow-men.

FIRTHS AND SEA-LOCHS

These words have no precise scientific meaning and are virtually interchangeable. The inlets of the sea on the east coast are almost without exception described as firths, whereas the relatively narrow, fiord-like inlets of the west coast are known as lochs. There is no consistency on this point, however, for Loch Fleet is in eastern Sutherland and a few of the larger inlets on the west, such as the Solway and the Clyde, are distinguished as firths. To complicate matters further, there is a distinctively Scottish word, 'kyle', which is used not only to indicate a narrow channel between an island and the mainland—as in the Kyles of Bute—but also to describe a sea-loch on the north coast, the Kyle of Tongue. There seems to be no valid reason for this latter use of the word which would serve better if restricted to a synonym for strait. On the west coast there is also a number of sounds, a term used to define a larger stretch of water than a strait: for example, the Sound of Mull and the Sound of Sleat.

There is a marked contrast between the west- and east-coast inlets. There are many more on the west, and they are divided into groups with marked trends: the Argyllshire lochs aligned south-west/north-east; those in western Inverness-shire west-east; and the northern group of Ross and Sutherland north-west/south-east. They are comparatively long and narrow and some bear a distinct resemblance to Norwegian fiords. Some controversy has taken place as to the reasons for the differences: they are probably the result of several factors including the type of rock and the position of the main watershed. Perhaps the principal reason is fracturing, caused by earth movement, combined with the gouging effect of glacial action accentuated by the narrow valleys.

North of the English Lakeland hills lies the Solway Firth: 1,000 square miles of muddy creeks, twisting channels, seaweed-strewn boulders and interminable sand-banks which stretch almost from shore to shore. The Solway has a character completely different from that of other Scottish waters: an immensity of sea and sky more nearly akin to the East Anglian creeks, though on an infinitely vaster scale. For the marine biologist, the hydrographer, the student of coastal physiography, the ornithologist and the plant ecologist, this firth has much to offer. The enormous quantities of silt, sand and mud create the impression that the estuary is holding a perpetual debate with itself as to whether it is land or sea and cannot quite make up its mind. Certainly this unsettled state exists in the mud-banks: they are essentially unstable, causing the channels frequently to change their course. Apart from the mud-flats there are scaurs—a local term for deposits of boulder clay—which are numerous but exposed only at low tide. In one or two places, traces of a submerged forest can be seen at extreme low tide.

The extensive sand-banks way out in the estuary have a varied fauna. Burrowing in the sand at the perimeter are the banded wedge-shell *Donax vittatus* with its brightly coloured polished surface, the rayed trough-shell *Mactra corallina* and the thick trough-shell *Spinula solida*. These are offshore creatures, but the walker along the shore might well find the empty shells. On the scaurs there are beds of the horse mussel *Modiolus modiolus*: the largest species in Britain, it is more of an offshore creature than is the common mussel *Mytilus edulis*, beds of which exist in a number of places. Sea-firs, crabs, lobsters, starfish and many other marine animals abound. The invading Australian barnacle *Elminius modestus* began colonising the Solway in 1953. It is now widespread and like its British relatives is proving itself a nuisance by fouling boat hulls.

Fishing has been a Solway activity at least since Mesolithic times, when hunters with spears harpooned the fish. The district has its own highly distinctive methods of catching both salmon and whitefish. Stake or paidle nets are fastened on to permanent poles which are sited about 7ft apart and are placed far out into the estuary. Remains of very ancient poles, which may date back to Roman times, have been discovered. An alternative method used for fishing the salmon is by the haaf net, haaf being an old Norse word for channel. The salmon seemingly prefer the shallow water and the fishermen wade out with a portable but cumbersome net fastened on to a wooden framework. Great expertise is required for this task. Boats from Annan have for many years trawled for shrimps and there is a small crab and lobster industry. Flounders are abundant and can be found even in small salt-marsh pools; and in recent years the numbers of plaice have been increasing along the northern shore in late summer.

For many people the attraction of the Solway lies in its birds and especially in the geese which from autumn to spring frequent the sand-banks and salt-marshes in huge flocks. The most important area for the geese is the north-east shore where on Mersehead Sands, Blackshaw and Priestside Banks barnacle, greylag and pink-feet congregate. Large numbers of duck including wigeon, pintail, merganser, scaup and scoter can be seen all along the firth from Wigtownshire to Dumfriesshire. Two notable localities for scaup are in Rough Firth and Carsethorn: at the latter I have watched them close inshore as late as mid-April. As might be expected, waders are numerous along the shoreline and all the common kinds occur.

The upper part of the Firth of Clyde is highly industrialised, but the broad waters of the outer estuary are well endowed with rich beds of seaweed, including oarweed and the edible carragheen *Chondrus crispus*. Another edible plant which is also known as carragheen is the small tufted alga *Gigartina stellata* and the lower shore is characterised by dense belts of this weed. Research has been carried out by the Scottish Marine Biological Association into the effect on marine life of the warm-water outflow from the nuclear power station at Hunterston. One noticeable effect discovered was that of the earlier breeding of certain animals such as the small crustacean *Urothoe brevicornis* and the netted dog-whelk *Nassarius brevicornis*. One of the small molluscs which is abundant in the Clyde is *Hydrobia ulvae*, whose favourite food is sea-lettuce; and it has been estimated that it reaches densities of over 36,000 individuals per square yard. Mud-flats at various places on the southern shore hold fair

Page 53 (*above*) The Isle of Arran across the Sound of Bute; (*below*) oyster-catcher and chick on roadside shingle in Banffshire

Page 54 (*above*) River Luineag, Inverness-shire, a typical dipper-haunted Highland stream, draining Loch Morlich; (*below*) lupins naturalised on shingle of the river Dee at Ballater

numbers of waders at migration times and in winter have varying numbers of wildfowl including eider, merganser and wigeon. It must be admitted, however, that bird-wise the Clyde cannot compare with the Solway. Various cetaceans are not infrequently observed in the lower part of the firth; the commonest is probably the porpoise and others are the whitebeaked dolphin and the killer and bottle-nosed whales.

The sea-lochs of the west coast often have a sill or bar across their entrance, indicative of glacial action. A well-known example is that of the Falls of Lora in the mouth of Loch Etive; and a very marked one is the sill across Loch Morar which is so high that it keeps the sea at bay and has transformed the loch into a freshwater one. In a number of lochs where the entrance is relatively narrow, tide-races occur as the sea is compressed into the bottle-neck. The combination of sheltered conditions and the cleansing effect of the flow enables a rich fauna to develop. Western lochs have been classified as either exposed or sheltered, but there is no rigid distinction between the two types.

The Scottish Marine Biological Association, now based at Oban, is carrying out extensive researches into the Argyllshire sea-lochs. Much work has been done on Loch Etive, which is an outstanding example of a 'sheltered loch'. The quantities of brown seaweed lessen towards the head of the loch; serrated wrack *Fucus serratus* and bladder-wrack *F. vesiculosus* are here found at a lower level than on the open coast. The lower shore has the oarweed *Laminaria digitata* in dense masses; but the specific name is not really apposite for the plants in this loch since they are of an atypical form. At the head of the loch is a brackish-water community with the wrack *F. ceranoides*, the sandhopper *Gammarus zaddachi* and several species of bivalve. A recent discovery is the presence in the loch of a sea-squirt of northern waters which is a new species for the Scottish list.

The types of fish found in these sheltered waters vary somewhat from loch to loch. Those which are commonly found include cod, herring, sprats, and flat-fish such as plaice, dab and flounders; off the coastal rocks saithe are often abundant. The last-named is sometimes known as coalfish, because of the dark colouration of its upper parts which varies from brown to bottle-green. It is essentially a northern species, becoming increasingly common in the far north. Although the older fish live in deep water off rocky shores, the younger fish prefer shallower water around coastal rocks and in the sea-lochs. These young fish are often known as sillocks, although they are accorded a number of other local names. Fishing for them in the summer months in north-west Scotland and the northern isles is a long-established tradition among the crofters. The White Fish Authority is carrying out experiments in farming saithe at Faery Island on Loch Sween.

Loch Etive has a varied bird population which can be regarded as representative of these western sea-lochs. Near the mouth of the loch the summer-nesting birds are rock-pipits, common and arctic terns and the ubiquitous eider. Oyster-catchers and common sandpipers breed along the loch shores and the red-breasted merganser is a regular nester. In winter such birds as the whooper swan, the three species of diver, golden-eye and wigeon frequent the loch, the last-named being the commonest wintering duck. A number of otters live along the loch and they are widespread along the whole of the west coast from Argyll to Sutherland. They are probably more plentiful along this western coast than anywhere else on the mainland of Britain.

The sea-lochs of north-west Scotland are relatively long and narrow and they have physiographic conditions which correspond to those of sheltered seacoasts. Not only are the wracks the dominant algae but they often grow to a great size. The seaweed of rocks and boulders at mid-tide level is the knotted wrack *Ascophyllum nodosum*. Shingle at this level has bladderwrack and in the zone below, serrated wrack. Of the marine snails, the rough periwinkle *Littorina saxatilis* is locally dominant; this is a variable species and the form usually found in these lochs is larger and lighter in colour than that found on exposed shores. The common periwinkle *L. littorea* is less widespread but occurs in places with the flat-top shell *Gibbula umbilicalis* although this last-named species becomes progressively less common northwards. As usual, the area at low-tide level and below where the waving fronds of oarweeds grow, holds the richest life. Although mussels are more typical of open coasts, dense beds are sited in some lochs: for example, in some Argyllshire lochs and in the Kyle of Durness. Common marine animals include the snakelocks anemone *Anemonia sulcata*, the grey-top shell *Gibbula cinerea* and the small, pale green sea-urchin *Psammechinus miliaris*.

Loch Eriboll on the north coast of Sutherland can be regarded as one of the more exposed sea-lochs. There are quantities of serrated wrack, a species with a wide-exposure range; bladderwrack, which on exposed coasts has developed a small bladderless form; and knotted wrack, which although plentiful along the west coast is only known from two lochs on the north coast. Mussels and barnacles are animals that are more at home in exposed conditions and both are found in quantity here. There are the usual summer-breeding birds and there is a possibility in summer of a glimpse of flighting greylag geese belonging to the small resident population in Sutherland. Generally, the more open lochs have a greater variety of birds although much depends on the nature of the shore; the Kyle of Tongue, in addition to its open character, has wide mud-flats which are attractive to waders on passage. Herons are common at all seasons in the north-west sea-lochs, and when there are no trees, they often nest at ground level on islands such as those in Loch Laxford.

Reference has been made earlier to the great differences in appearance between the west- and east-coast inlets and these differences are reflected in the plant and animal life. The wracks are much less abundant in the eastern firths, although the Firth of Tay is fairly well endowed not only with wracks but with various green seaweeds. The marine life generally is not as rich as in the more rocky western lochs, although some species are numerous. The vast expanse of sandy mud-flats contains lugworms *Arenicola* spp, catworms *Nephthys* sp, shrimps *Crangon vulgaris* and cockles *Cardium* spp. Along the shores of the Firth of Forth are the beadlet anemone *Actinia equina* and the dahlia anemone *Tealia felina*; the Australian barnacle in its northward spread reached the Forth in 1958.

As might be expected, these firths with their muddy shores are much superior to the sea-lochs in ornithological interest, both in variety of species and in sheer weight of numbers. Winter is the time when large—in some places immense—flocks of wild-fowl resort to these waters. Pink-footed geese frequent most of the principal inlets from the Tyne estuary at Dunbar in the south to Dornoch Firth in the north; their numbers are at their greatest in the autumn and spring migrations, since many move farther south. Increased nesting success in Greenland and extension of breeding range since just before the beginning of the present century have resulted in much larger numbers on the east coast in recent years. The upper part of the Tay firth is a particularly favoured spot, where there may be 20,000 or more in the autumn. Greylags winter in nearly all those waters occupied by pink-feet, but often in smaller numbers. Thousands of sea-duck, common and velvet scoter, red-breasted merganser, goosander, golden-eye, scaup, pintail, long-tailed duck and eider congregate in the firths during the winter months although no one particular water is likely to have all these species. Autumn is naturally the best season for waders and at this time many uncommon migrants occasionally arrive, even North American birds such as the Wilson's phalarope in the Firth of Forth in 1954: the first record for Europe. This was regarded as a freak occurrence, but since then the species has been seen a number of times. In fact, by the end of 1971, there were forty-seven records for Britain and Ireland; and it looks as though this bird may become one of the fairly regular transatlantic migrants which arrive in Britain in very small numbers when weather conditions favour the crossing. The usual waders occur, but black-tailed godwits are not plentiful although they frequent the estuary of the Eden. There are sizable flocks of bar-tailed godwits, especially in the Firth of Tay. Where waders en masse are concerned it is the knot which must hold the centre of the stage, for on the mud-banks of the east-coast waters and along the Solway shores in the west, vast flocks of knot congregate in winter.

ROCKY SHORES

The rocky shore is the basic coastal habitat in Scotland; most prevalent in the west and north, it has still a wide distribution along the eastern seaboard. Dr J. R. Lewis has made an intensive study of the plant and animal communities of the rocky western coastline, and the results have been described in several contributions to the Transactions of the Royal Society of Edinburgh. His book, *The Ecology of Rocky Shores*, also includes a fine and detailed analysis of the west Scottish coast to which the reader wishing to pursue this fascinating subject in depth is referred.

My own first acquaintance with the plant zonation on coastal rocks of the west was at Traigh Sands, Arisaig. Here I found that the dark slabs of Schistose Grit veined with quartz were made decorative with orange and grey tufted lichens. The former was *Xanthoria parietina* and the latter *Ramalina* sp. Below this was a dense belt of the blackish-green channel-wrack *Pelvetia canaliculata*. Under this was a parallel belt of an olive-brown seaweed, the flat wrack *Fucus spiralis*. At this level on the shore this was the lowest-growing seaweed, though in the small rock-pools the thin green ribbons of *Enteromorpha* sp were plentiful. On the bare rock face between the masses of wrack were barnacles and limpets. The tiny shells of the small periwinkle *Littorina neritoides* were scattered in the crevices and upper rock surfaces, whilst below them were numbers of flat and common periwinkles.

Such was the relatively simple picture of the rocks as seen by a casual stroller on the beach. I know now that in reality the scene is very much more complex. The zonation is worth looking at in a little more detail, but it cannot be too strongly stressed that there is no rigidity in the vertical distribution and some of the marine life may be found occupying a wider range than the particular zone to which they have been allotted below. The highest part is the splash zone or littoral fringe as some marine biologists prefer to designate it. The plants here are lichens and on these open western coasts exposed to the Atlantic this zone occupies a greater depth than in the sea-lochs. Below the orange *Xanthoria* is a belt of black encrusted lichen *Verrucaria maura* which has a tarry appearance and does in fact stain the fingers when it is handled. With it or just below on steep slopes may be one of two *Lichini* spp. Animals here are the variously coloured small periwinkle, the sea-slater *Ligia oceanica*, a marine bristle-tail *Petrobius maritimus* and various mites. Small pools at this level will have the seaweed *Enteromorpha* with the rough periwinkle and the ever-present shore-crab *Carcinus maenas.*

Proceeding seawards we come to the upper littoral zone with the successive belts of channel- and flat-wracks mentioned earlier. Red seaweeds are usually associated with the lower shore but the purple-coloured and edible laverbread *Porphyra umbilicalis*, on the exposed coasts of both west and east, extends upwards almost to the level of the channel-wrack. The rough periwinkle is the typical member of its genus at this level and the small periwinkle extends downwards to mingle with the former species. Acorn barnacles now come into the picture and it is clear that in exposed situations they face competition from the common mussel, which at mid-shore level is often present in dense beds. Despite this, however, barnacles are both widespread and exceedingly abundant, especially on steep rock surfaces along the Scottish coasts. These barnacle-encrusted rocks are very conspicuous and from a distance they look like a horizontal band of chalk. H. Barnes and R. L. Stone report that the Australian barnacle has now reached Dunstaffnage Bay near Oban on the west coast and the Firth of Tay on the east; these localities are the farthest north in Europe at which the species has so far been recorded.

The mid-shore contains a great variety of plant and animal life. Where conditions permit the growth of seaweed, bladderwrack forms the layer below flat wrack and largely replaces the knotted wrack which is typical of more protected shores. The saw-wrack follows below and is common except in conditions of extreme exposure. A number of red and green seaweeds grow here, including the green sea-lettuce *Ulva lactuca*, which is reputed to be a favourite food of limpets. The common limpet *Patella vulgata* lives at the higher levels of the mid-shore and other species occur below this. The barnacles extend downwards into this zone and are preyed upon by the dog-whelk *Nucella lapillus*. Other molluscs include the common periwinkle and the flat and grey-top shells. The typical anemone of the mid-shore is the beadlet *Actinia equina*.

On rocky shores between low and high water-mark on all the Scottish coasts, there are innumerable rock-pools either on the open, gently sloping platforms or in hollows surrounded by massive boulders. For the lover of nature, there can be few more delightful ways of spending an hour or so at the seaside than by lying face down on a sun-warmed rock and watching with fascinated eyes the teeming life and exotic beauty clearly visible in the water a few inches below one's face; while always in the background is the rhythmic roar of the sea. The pools themselves are almost microcosms of the sea, though without the action of the waves.

The plants and animals which the pools contain must withstand great variations in the physical conditions: for example, in the temperature, salinity and in the amount of desiccation to which they are exposed. A variety of seaweeds thrive in the rock-pools. The shallow ones in particular have small, delicate green, pink and red seaweeds. Some of the commonest are the exquisite coral-pink tufts of *Corallina officinalis* and the flat encrustations of the very similar *Lithophyllum* and *Lithothamnion* spp which coat the rock surface with pastel shades of purple-pink. Then there are the green weeds, the dark wiry tufts of *Cladophora rupestris*, the long green ribbons of *Enteromorpha* and the flat leaves of the sea-lettuce. These small shallow pools are miniature underwater rock-gardens, not a whit less beautiful because there are no flowers. In the deeper pools one has the opportunity of seeing at close quarters the marine life of that otherwise largely inaccessible zone of the sub-littoral below low water-mark.

If the plant life is rich, so also is the animal life. Limpets, periwinkles and top shells decorate the floor and walls of the pool at random and there are occasionally dog-whelks and common whelks as well. Shrimps and prawns propel themselves through the water, crabs make their ungainly way across the floor and jellyfish are sometimes left stranded by the tide. The starfish is another common inhabitant and those colourful animals, the anemones, are conspicuous especially in pools on the west coast. They include the beadlet, snakelocks, plumose *Metridium senile*, *Sagartia elegans*, *S. troglodytes* and *Cereus pedunculatus*. There is considerable colour variation in anemones and unfortunately some of them are difficult to identify, but this does not detract from one's aesthetic appreciation of them. This list does not exhaust the faunal content of the pools; there are also the small fishes which are so characteristic of this habitat, darting from one weed-covered recess to another. The three-spined stickleback is one of them: a surprising fish this, which I have found in habitats as geographically and as ecologically remote from each other as the chalk rivers of Hampshire and the rock-pools of the Shetlands. The eel-like butterfish is another, and this species is reported to figure largely in the diet of the otter. The common blenny, the various gobies and the fifteen-spined stickleback are fairly common, while wrasses and sea-scorpions are occasional. In the Firth of Lorne a common blenny has been recently discovered with an associated leech *Oceanobdella blenii*, a southern species which has not been known in Scotland before. The viviparous blenny lives in the rock-pools along the Clyde and is probably generally common; but its relative, Yarrell's blenny, is much more scarce.

The lower shore from low water of neap tides downwards has, too, its distinctive flora and fauna. Here grow the large brown seaweeds, the thongweed *Himanthalia elongata* and, just below, the long waving fronds of *Alaria esculenta* with its conspicuous mid-rib, a species typical of exposed coasts. Where wave action is less severe, *Laminaria digitata* may replace *Alaria*. Certain small red algae grow on the *Laminaria* and three of them are of northern distribution. *Ptilota plumosa* is an attractive plant with feathery branches: it is quite common in Scotland, as is *Petrocelis hennedyi*, named after the Scottish botanist R. Hennedy; and the third is *Rhodomela lycopodioides*. Apart from these, most of the free-living red seaweeds are found in this zone. One of them, *Odonthalia dentata*, is essentially a northern species which is generally common on Scottish coasts; and among the others is the pepper dulse *Laurencia pinnatifida* which is locally plentiful. The downward extension of mid-littoral weeds into the lower shore can be illustrated by the small yellow-brown seaweed *Mesogloia vermiculata*, a common species which is particularly abundant in north-east Scotland. Mid-shore molluscs which extend into the lower zone are the grey-top shell and the limpet *Patella aspera*; in some sheltered places the tortoiseshell limpet *Acmaea tessulata*—a northern species—and the white tortoiseshell limpet *A. virginea* may sometimes be found.

Herring gulls have increased in numbers in Scotland and are plentiful breeders in this situation. The great black-backed gull nests in smaller numbers on stacks and islets but also occasionally inland; the common gull on the other hand is principally an inland nester but breeds also on coastal stacks. The common birds of rocky coasts are rock-pipits, oyster-catchers and eiders. In mild winters, stonechats, instead of migrating south, have moved to the coastline where they have been recorded in recent years on the Banffshire coast. Turnstones and purple sandpipers are common winter visitors and some non-breeding birds of the former species remain throughout the summer in the north-west.

CLIFFS

Cliffs of varying lithological character line the shores of Scotland and there are raised beaches in many places on all coasts.

Caves exist in some localities, such as on the Kintyre peninsula, in the low cliffs of New Red Sandstone on the Moray coast, in Old Red Sandstone on the Angus coast and in the Cambrian Limestone at Durness where is the well-known Smoo Cave.

As a habitat for plants, cliffs vary greatly. They may consist of steep grassy slopes, tumbled boulders in a sheltered bay or a near-vertical face rising sheer from the sea on exposed headlands. The sheltered spots naturally have the most luxuriant vegetation, which is not restricted to purely maritime plants. The sandstones and lavas of the Devonian period at various places along the eastern seaboard are sufficiently basic in type for lime-loving plants to obtain a foothold. So the blood-red cranesbill *Geranium sanguineum*—which the plant-lover will associate with the limestone of the Gower peninsula in Wales—flowers on these cliffs, as does the purple milk-vetch *Astragalus danicus*, the carline thistle *Carlina vulgaris* and the clustered bell-flower *Campanula glomeratus*, all plants of English downland. Another plant of calcareous soil which is more typical of sea-cliffs, is the kidney-vetch or, as it is sometimes known, lady's fingers, *Anthyllis vulneraria*. Bird's-foot trefoil *Lotus corniculatus* makes splashes of bright yellow on the cliffs during the summer months. Where the strata are of an impermeable nature, the meadow-sweet *Filipendula ulmaria*—a plant of streamsides—is able to grow on the moist ledges. Drier ledges will have viper's bugloss *Echium vulgare* and the stonecrops *Sedum acre*, *S. album* and *S. anglicum*. The first two stonecrops are characteristic of basic soils whilst the last-named is typical of acid granitic rocks. Along some sheltered cliffs bracken flourishes and there are woodland plants such as primrose *Primula vulgaris*, wood-sage *Teucrium scorodonia*, red campion *Silene dioica* and wood-vetch *Vicia sylvatica*. Plants more especially associated with sea-cliffs are thrift *Armeria maritima*, sea-campion *Silene maritima*, scurvy-grass *Cochlearia officinalis*, sea-plantain *Plantago maritima* on moist rocks and buck's-horn plantain in drier places. The fern, sea-spleenwort *Asplenium marinum* is a northern species which grows in rock crevices along the eastern coast, although it is absent from some localities. Another characteristically Scottish plant is lovage *Ligusticum scoticum*, an umbellifer which is locally frequent on rocks all round the coasts. The cliff form of the Nottingham catchfly *Silene nutans* var *smithiana* grows on the Kincardine cliffs.

The exposed coastal cliffs of the north-west are splashed with the vivid colours of lichens and in the crevices grow thrift, English stonecrop and rose-root *Sedum roseum*. Here and there in the short turf on the cliff-tops along the western coast in spring the soft blue flowers of the vernal squill *Scilla verna* may be found and where it does grow it is abundant. Spreading over acid rocks the dwarf shrub crowberry *Empetrum nigrum* opens its small pink flowers in early summer.

The cliff flora of the northern coastline is of especial interest for two reasons. One is that this is the only part of the Scottish mainland where the Scottish primrose *Primula scotica* grows. Anyone who thinks of this plant in terms of the common primrose is likely to have some difficulty in finding it, for it is minute in size and almost hidden even by the short turf in which it grows; but once it has been located other plants will soon be discovered near-by. The flowers are purple-red and altogether it is an attractive little plant. It is not confined to cliff grassland but grows also on dunes and on inland pasture. Plenty of moisture is an essential requirement.

The other botanical interest of this coast lies in the presence of plants of the high mountains growing here at near sea-level. The reason for this is simply that, strange as it may appear at first sight, the conditions in the two habitats are very similar. Both are open exposed sites. Both suffer physiological drought during the winter by reason of the shallow nature of the soil and by their not being able to take up the available rain. Perhaps the most conspicuous of these mountain plants is the lime-loving mountain avens *Dryas octopetala* which drapes the rocky outcrops with its pale, creamy white flowers in June. The moss campion *Silene acaulis*, the yellow mountain saxifrage *Saxifraga aizoides*, the mountain sorrel *Oxyria digyna* and the purple saxifrage *S. oppositifolia* are other alpine species found, sometimes in abundance, on the northern coast. At the other end of the scale is the purple mountain milk-vetch *Oxytropis halleri* which may be considered a maritime plant but is one which also occurs, though rarely, at a high altitude.

Scotland does not possess anything like the variety of insects found in the southern half of Britain; for example, of some sixty-seven species of butterflies on the British list only thirty-one occur in Scotland. One of the best places to see a good selection of butterflies is on the well-vegetated slopes of the eastern coastline where in places the common blue *Polyommatus icarus*, small blue *Cupido minimus*, meadow-brown *Maniola jurtina*, small heath *Coenonympha pamphilus*, grayling *Hipparchia semele*, green-veined white *Pieris napi* and dark green fritillary *Argynnis aglaia* fly in the sunshine.

Certain moths frequent the coast and the southern part of the country is likely to be more rewarding of these since the number of species declines gradually as one travels northwards. The marbled coronet *Hadena confusa* is a beautiful little moth with variegated forewings; its larvae feed on sea-campion. On the east coast the typical form is replaced by one with darker colouration. The small elephant hawk moth *Deilephila porcellus* is not restricted to the coast, but it does have a preference for a maritime environment and occurs in a few places on the east coast of the Southern Uplands. Why it should have this preference is by no means clear, since its caterpillars feed on a variety of common plants. The grey chi *Antitype chi* is also found on the coast as well as inland, the determining factor here probably being the open nature of the habitat. The six-spot burnet *Zygaena filipendula* is not uncommon and two rare burnets *Z. purpuralis* and *Z. achillae* occur in one or two localities on the west coast, in Argyllshire.

Sea-bird communities consisting of common guillemots and razorbills (sometimes in very large numbers), puffins (generally in much smaller numbers), fulmars and kittiwakes occupy various headlands; but common guillemots are unaccountably scarce on the western mainland. Two races of this species breed in Scotland. The vast majority are of the northern race, which has a black back; but the southern race, with dark chocolate-brown upper parts, is found on the mainland on the cliffs of Galloway. Baxter and Rintoul record an amusing incident of a guillemot's travelling at an air-speed of 45 mph accompanying, and over a distance of several miles slowly overhauling, a destroyer on speed trials. The auk populations have been reduced in recent years by large-scale wrecks such as that in the Irish Sea in 1969, when over a period of some eight weeks about 50,000 birds died. Whilst many recent sea-bird deaths can be attributed to pollution, it must be admitted that the decline in the puffin population began in the last century in southern England and extended to western Britain in the present one, so that this long-term decrease may be affecting northern colonies now and contributing an additionally adverse factor. The reasons for this long-continued decline are not known with certainty, but one suspects predation by that aggressively successful species, the great black-backed gull.

The black guillemot is of quite different character from its larger relative, and lacks the gregariousness of its kind, although it will tolerate small numbers of its own species nesting in proximity provided that they do not invade its privacy. The nest site is often an individual one in an isolated crevice, unlike those of the long, overcrowded ledges of the colonial sea-birds. In Britain the black guillemot is restricted chiefly to the Scottish coasts where it is widely if somewhat locally distributed. To the Highlander it is known as the tystie but in the Middle Ages the early naturalists called it the sea turtle-dove. Fulmar petrels first began colonising the north Scottish mainland about the beginning of this century and have gradually spread southwards on both the west and east coasts. On the east coast, St Abb's Head in Berwickshire was occupied in 1921; but on the west coast the first breeding on the Mull of Galloway was not until 1932. Today the fulmar is a common bird which occupies its nesting sites for much longer in the year than do the auks, leaving the cliffs in late autumn for a few weeks far out on the ocean. Kittiwakes must rank as the most beautiful of British breeding gulls and it is good to know that they are on the increase, nesting in many places on the cliffs. Herring gulls have also increased but they are not restricted to ledges and will nest low on the shore.

Shags are commoner than cormorants in most areas, but along the western coasts of the Southern Uplands the cormorant may be greater in numbers than the shag. Five members of the crow family are cliff breeders in Scotland. The raven, though not confined to coastal cliffs, is more plentiful there and it is a fairly common bird on the west and north coasts, much less so on the east. The chough was once a regular breeder on the mainland and it was plentiful up to 1835, but in the next thirty years it virtually disappeared. Isolated cases of nesting were reported up to 1937 and perhaps later; and there remains a faint possibility that the chough may once again become established on the mainland in small numbers. Jackdaws are often found nesting on the cliffs but they are less common in northern Scotland. Hooded and carrion crows are the remaining members of the family which nest wherever they are allowed to do so. The golden eagle and the peregrine falcon have coastal nesting sites in some places. D. A. Ratcliffe's 1971 survey of the peregrine population in Britain revealed that the breeding status on Scottish coasts has in most places deteriorated still further since the earlier survey carried out in 1962; and this may well be attributed to pollution of the sea, for these coastal peregrines feed on sea-birds. The sea-caves are the haunt of rock-doves and in some of them house-martins and kittiwakes also nest.

CHAPTER FOUR

The coastline continued

Mud-flats–Shingle beaches–Sandy beaches–Sand-dunes–Salt-marshes–Coastal waters

MUD-FLATS

EXTENSIVE MUD-FLATS LINE the firths of Solway and Clyde in the south-west, the Kyle of Tongue on the north coast and the inlets of the east. This is a very specialised environment requiring of its plants both salt-tolerance and ability to withstand the alternations of the tide. Apart from the salt-marsh communities the flowering plants in this situation are the three grass-wracks *Zostera* spp and these live, flower and pollinate under the sea. The common grass-wrack *Z. marina* grows at the lowest level and the other two between the tides. They are the preferred food of brent geese but about forty years ago they were attacked by a disease which decimated the population. They are now of very local distribution in Scotland with *Z. marina* probably the most widespread.

Many of these muddy shores have a sand content as well, and they possess a characteristic fauna. Lugworms *Arenicola marina* are abundant and sometimes associated with them is another member of the bristle-worm family, the catworm *Nephthys* spp. Two common sand-hoppers are *Bathyporeia pelagica* and *Corophium volutator*. In a study of the redshank on the Ythan estuary, Dr J. Goss-Custard found that *Corophium* was the most important prey; it has very high densities in unpolluted waters and it has been estimated that there are over 25,000 per square yard in the Ythan. Other crustaceans include the common shrimp *Crangon vulgaris* and various crabs. Two molluscs which are particularly associated with the brackish water of the firths are the peppery furrow-shell *Scrobicularia plana* and the Baltic tellin *Macoma balthica*; they are recorded from the Solway and are probably widely distributed in similar situations. Two edible bivalves of sandy mud found on a number of Scottish shores are the sand-gaper *Mya arenaria* and the common cockle *Cardium edule*. Another edible species which, although more characteristic of rocky coasts, is also found on the muddy shores of the firths, is the common mussel *Mytilus edulis*. The common whelk *Buccinum undatum* too is widespread.

SHINGLE BEACHES

Beaches consisting of shingle or part shingle and sand are numerous; they exist along the Solway and Clyde shores, in parts of the west coast, on the Nairn and Moray coastline and in Fifeshire, among other places. In contrast to the beaches of southern England, with their abundance of flint pebbles, the Scottish beaches are largely comprised of pebbles from the igneous and metamorphic rocks. Siliceous pebbles on the Ayrshire coast sometimes contain agates which are known as Scotch pebbles and are used in jewellery.

There is naturally a certain amount of variety in the composition of the flora, but all the plants in one way or another are able to adapt themselves to this specialised environment with its conditions of drought, salinity and unstable substratum. Sea-campion *Silene maritima*, Babington's orache *Atriplex glabriuscula*, scentless mayweed *Tripleurospermum* and curled dock *Rumex crispa* are often co-dominants. In places silverweed *Potentilla anserina* is very plentiful and cleavers *Galium aparine*, sorrel *Rumex acetosa* and oat-grass *Arrhenatherum elatius* are other common plants. Cliff plants such as thrift *Armeria maritima* and scurvy-grass *Cochlearia officinalis* are sometimes found on the shingle as well as sea-purslane *Honkenya peploides*, which is more typical of sandy shores. A plant of northern distribution, the oyster-plant *Mertensia maritima* is dotted sparingly along the

Page 71 (*above*) Heron fishing the upper reaches of the Tweed in Berwickshire; (*below*) globe-flower, a northern plant of moist habitats

Page 72 (*above*) Visitors queue to see the ospreys at the Loch of the Lowes, Perthshire; (*below*) Slavonian grebe among horse-tail on an Inverness-shire loch

west coast from the Solway Firth northwards, along the northern seaboard and on the southern shores of the Moray Firth. In the last forty years it has experienced a marked contraction of its range. This procumbent member of the Borage family is named after an eighteenth-century continental botanist Franz Karl Mertens; it has glaucous fleshy leaves and flowers which are pink at first, changing to purple-blue.

Marine animals are very scarce on shingle. A bristle-tail *Petrobius maritimus* occurs under stones chiefly on the west coast and another member of the genus *P. brevistylis* is found associated with large boulders principally on the east coast. The sea-slater *Ligia oceanica* is a relatively large marine wood-louse which is often very plentiful in shingle. Under the larger stones, various encrusting organisms including barnacles and sea-squirts will be found.

Birds of shingle are also few. Common breeding birds whose preferred nesting site is shingle are the ringed plover and the oyster-catcher. The former is wonderfully camouflaged in its pebble surroundings, even on purple Torridonian Sandstone pebbles. Terns nest in various situations including shingle, although they are perhaps most often found on the offshore islands. Out of the breeding season, parties of turnstones forage among the stones and in recent years a few shore larks have begun to winter along the east coast.

SANDY BEACHES

Mention has already been made of the various kinds of fine sandy beach, many of them in superb settings, which abound along the Scottish coasts. Despite the ever-increasing numbers of visitors, some of the best of these beaches are virtually untrodden even at the height of the holiday season. Relatively few flowers grow on sandy shores, in contrast to the great variety found on sand-dunes, and almost all of them are annuals. They are thus able to avoid the severity of the winter gales by surviving in the form of seeds, although in conditions of extreme exposure even annuals may not be able to grow. Several kinds of orache are to be found near the high-tide mark, together with sea-rocket *Cakile maritima*, saltwort *Salsola kali* and the sea-purslane which, as noted already, is more plentiful on sand than on shingle. The Scottish scurvy-grass *Cochlearia scotica* grows principally in northern Scotland on sand, shingle and in rock crevices.

Many molluscs including razor-shells, cockles and tellins burrow in sandy shores. The thin tellin *Tellina tenuis* is typical of clean sand from mid-shore downwards and *T. fabula* from lower levels still. Exposed beaches have the banded wedge-shell *Donax vittatus*. Although sand-hoppers are neither restricted to sand nor most plentiful there, some, such as *Talitrus saltator*, are numerous in this habitat. Shrimps *Crangon vulgaris* are common at the lower levels and a very small crustacean *Asellopsis intermedia* is recorded as abundant on the sandy beaches of the Clyde. At the top of the beaches the strand-line of organic debris is a zone where land and marine animals overlap and it has its own fauna composed chiefly of small grey flies, sand-hoppers and many kinds of beetle.

The shallow waters offshore have certain fish which are especially associated with sand. There are five species of sand-eel, those silver-coloured fish which the visitor to the shore so often sees, several at a time all neatly lined up and facing the same way, in the beaks of puffins. The various species live at different levels and the lesser sand-eel is a true shore fish living between the tides. The young of certain flat-fish, notably plaice, dab and lemon sole are abundant in the shallow sandy bays all round the Scottish coastline, the last-named more especially off western and northern shores. An astonishing anatomical feature of flat-fish is the movement of one of the eyes. At first the eyes are in the normal position and the young fish swims vertically, but soon one eye moves to join its fellow in readiness for the adult stage when the fish swims on its side. Some flat-fish have eyes on the right side whilst others have them on the left; the species mentioned above, being members of the plaice family, have the eye on the right. Another physical development resulting from the change to swimming on the side is that the colour on the underside is lost and the top side often develops a mottled pattern which makes an excellent camouflage in the natural habitat. One factor which can help in distinguishing the various species is the number of rays in the tail, and I once spent part of a wet morning on the island of Handa struggling to count the tail-rays of a very tiny plaice caught in a rock-pool.

A significant development in recent years which may well make a valuable contribution to Scottish fisheries is the establishment of fish-farms on the west coast. The experiment began in Loch Sween, Argyllshire during World War II when increased food production was essential. The water was artificially enriched with minerals and the result was a much more rapid growth of flat-fish. In 1965 a plaice-farm was established in a creek at Ardtoe, on the Ardnamurchan peninsula, where young fish are brought on by artificial feeding in specially constructed tanks. I am informed that by this method plaice reach ½lb in two years. Recently the experiments have been extended to include another flat-fish, the turbot, which is less common than plaice in the natural state in Scotland but of high commercial value. The turbot family differs from that of the plaice by having the eyes on the left side.

SAND-DUNES

A number of fine dune formations occur on all coasts. Where the dunes have long been stabilised a calcareous grassland has formed from a basis of wind-blown shell-sand; but for the fullest development of the machair, as it is known, one must go to the islands of the Hebrides. A considerable area of dunes has evolved at Invernaver on the north coast, whilst really extensive dunes are found at a number of places on the east coast such as on the shores of the Moray Firth—notably the Culbin Sands—the Sands of Forvie in Aberdeenshire, St Cyrus in Kincardineshire, Tentsmuir in Fife and Aberlady Bay on the Firth of

Forth. A characteristic of dunes is their tendency to migrate in line with the wind direction, often, though not necessarily always, with an onshore wind. This movement of coastal sand is seen at its most dramatic in times of severe gale, when many acres of land may be buried. The blanketing action happened on an extensive scale in the Culbin area of Morayshire, over a number of years in the sixteenth and seventeenth centuries. It was aided by the ill-advised removal of marram grass by local people. It culminated in a night of tempestuous storm in 1694, when the large estate lying to the east of the sand-hills with its dwellings, farm-buildings and fertile land was buried under a thick carpet of sand. Similar movements of sand had been taking place at Forvie in Aberdeenshire over many years, achieving its devasting climax with the destruction of Forvie village about twenty years before the Culbin disaster. It must have been the latter which finally triggered off government action, for in 1695, during the reign of William III, the Scottish Parliament at Edinburgh passed an Act prohibiting the removal of marram from the dunes.

The initial colonising plants are the same as elsewhere in Britain, that is, sand-couch *Agropyron junceiforme* followed first by marram *Ammophila arenaria* and to a lesser extent sea lyme-grass *Elymus arenarius*, then later by the sand-fescue *Festuca rubra* var *arenaria*. Once these plants have succeeded in stabilising the sand, a considerable variety of wild flowers finds a congenial home in this habitat. Maritime plants include sea-holly *Eryngium maritimum*, sea-plantain *Plantago maritima* and sea-stork's-bill *Erodium maritimum*. Many are flowers which, although found in several habitats, are especially typical of dry waste places. Such are ragwort *Senecio jacobaeus* which is an early coloniser, lady's-bedstraw *Galium verum*, milfoil *Achillea millefolium* and bird's-foot trefoil *Lotus corniculatus*, whose red-tipped yellow flowers in June carpet the dune pasture behind Achmelvich beach on the Stoer peninsula. Orchids frequently seen on western dunes are the early purple *Orchis mascula*, the common spotted *Dactylorchis fuchsii*, the fragrant *Gymnadenia conopsea* and the common tway-blade *Listera ovata*. The sand meadow-rue *Thalictrum minus arenarium* is a dominant species in parts of the St Cyrus dunes. The early sand-grass *Mibora minima* is a very rare plant of coastal sand which had not been known in Scotland for many years, but in 1961 it was rediscovered on sand-dunes on the southern shore of the Firth of Forth.

Often the soil is calcareous, particularly in the early stages. Lime-loving plants include kidney-vetch *Anthyllis vulneraria*, rest-harrow *Ononis repens*, blood-red cranesbill *Geranium sanguineum*, frog-orchid *Coeloglossum viride* and, along the east coast, the purple milk-vetch *Astragalus danicus*. The slacks between the ridges are frequently near the level of the water table. In consequence they have a number of moisture-loving species which are often lime-lovers as well. The sand-dune form of the early marsh-orchid *Dactylorchis incarnata* var *coccinea* has a much deeper colouring than the flesh colour of the type; and the dwarf purple orchid *Orchis purpurella* is another beautiful little orchid which is a feature of some of the west coast dunes. The frog-orchid is of catholic habit and grows on the slacks as well as on the drier slopes. The coral-root *Corallorhiza trifida* differs from the other orchids so far mentioned in that it is saprophytic, that is, it is a plant which is unable to manufacture its own food and has to obtain nourishment ready-made from organic matter in the soil. This is an uncommon orchid which grows in such widely differing environments as the shade of pine woods and exposed dune slacks on the east coast. The jointed rush *Juncus articulatus*, self-heal *Prunella vulgaris*, grass-of-Parnassus *Parnassia palustris* and penny-wort *Hydrocotyle vulgaris* are common members of the dune-slack flora.

At a late stage in the succession, acidic heath sometimes develops on the inner dunes with an ericaceous community of ling *Calluna vulgaris*, bell-heather *Erica cineria* and crowberry *Empetrum nigrum*. Non-ericaceous plants include the heath spotted orchid *Dactylorchis maculata* var *ericetorum*, sand-sedge *Carex arenaria* and the uncommon shepherd's cress *Teesdalia nudicaulis* at Tentsmuir. There is a considerable number of moss and lichen species on these dune heaths of the north-east. On mature dunes a few shrubs appear: hawthorn *Crataegus monogyna*, sea-buckthorn *Hippophae rhamnoides* and eared sallow *Salix aurita*, dwarf willow *S. repens*, alder *Alnus glutinosa* and the attractive white-flowered Scots or burnet rose *Rosa pimpinellifolia*. Whilst the majority of the wild flowers of the Scottish dunes are typical of this habitat throughout the British Isles, there are a few which belong essentially to a northern element in the flora. These include the globe-flower *Trollius europaeus*, the purple mountain milk-vetch *Oxytropis halleri*, the dark red helleborine *Epipactis atrorubens* on calcareous sand at a few places on the north coast and the sea-centaury *Centaurium littorale* which grows at Culbin and Tentsmuir.

The characteristic moths of coastal sand-hills are much the same throughout Britain, but where Scotland is concerned there is a slight variation from the general pattern. The sand dart *Agrotis ripae* is distinctly rare; the oblique-striped *Mesotype virgata* does not occur at all; the shore wainscot *Mythimna litoralis* is restricted in the main to the coasts of the Southern Uplands and the micro *Anerastia lotella* to south-west Scotland. By way of compensation the coast dart *Euxoa cursoria* is more of a northern species and occurs more plentifully along the east coast of Scotland than in England. Other moths of the dunes which are fairly well distributed in Scotland are the archer's dart *Agrotis vestigialis*, the white-line dart *Euxoa tritici* and the Portland moth *Ochropleura praecox*. In the butterflies the grayling *Hipparchia semele* is especially typical of Scottish dunes and the small heath *Coenonympha pamphilus*, meadow-brown *Maniola jurtina*, brown argus *Aricia agestis* and common blue *Polyommatus icarus* are also at home among the sand-hills.

The beetles of dunes are mainly of three kinds: the ground-beetles, leaf-beetles and the weevils. Although, naturally, there is not the rich variety as in the more congenial climate farther south, a number of species are found and I have seen that common insect of the English downland, the bloody-nosed beetle *Timarcha* sp on dunes only a few miles south of Cape Wrath.

Occasionally one finds an insect which is more plentiful in northern latitudes: one such on coastal sand-hills is the heath assassin bug *Coranus subapterus*. Certain spiders belong to this habitat: they include the small *Xerolycosa miniata*, a very local species, and the rather larger *Tebellus oblongus* recorded from dunes on the Morayshire coast. It is interesting to notice that both have the mottled sandy colouration which perfectly matches their surroundings.

A few land birds can be seen on the inner dunes. There are skylarks, yellow-hammers, linnets, stonechats and, if there is a good growth of vegetation, wrens. In winter, snow-buntings are found all along the east-coast dunes. Of sea-birds, up to four species of tern nest in places and black-headed gulls, shelduck, ringed plover and eider are regular breeding birds although the shelduck is scarce in the north-west. The Ythan estuary has the largest breeding colony of eiders in Britain. There are many rabbits and a black form is fairly common in parts of Scotland.

On a warm summer day with little wind to scatter the loose sand-grains, the hollows in the dunes are delightful places to be in, for the sand radiates heat, the ridges give shelter and the flowers are alive with visiting butterflies and bees. It is appropriate to end this section with a description of one of the well-known sand-hill areas—Aberlady Bay on the Firth of Forth—as it appears on a day in early August. There are mature dunes exceptionally well clothed with vegetation and possessing a rich fauna. A footbridge spans the Peffer burn and the path forks left along the western edge of the dune pasture, skirting the salt-marsh lining the creek. The industries and docks of the Edinburgh foreshore are only a few miles away to the west, but here all is peaceful and under the floating turrets of cumuli a sense of spaciousness prevails.

The blue flowers of viper's bugloss with their protruding pink stamens are plentiful along the path and act as a magnet to scores of insects. Butterflies are taking advantage of the sunshine: small heaths, common blues, green-veined whites, small tortoiseshells and a migrant painted lady. A little farther on, the leaves of a ragwort are being stripped by cinnabar moth caterpillars. Away on the right the bushes of sea-buckthorn not only give height to the scene but add an exquisite touch of colour with their slender, silvery grey leaves and their vivid orange berries which, a few weeks later, will provide welcome food for fieldfares and redwings as they flight in from Scandinavia. The sandy soil is moist in places and sustains a rich plant life of ragged robin, meadowsweet and marsh lousewort.

The path leads gently on to a sandy shore which earlier in the summer was populated by nesting terns. Now the site is bare, but the terns have not yet moved south, for a number of them are fishing offshore. With the resident ringed plover running along the strand-line are several migrating sanderling, elegant waders these but faintly humorous as they scurry along the sand with mechanical precision. Marram grass tops the dunes behind the shore until the rocky outcrop of Gullane Point is reached. From these rocks a large raft of over 300 eider can be seen floating lazily in the sea as they pass through their moult. A short distance inland is a small area of rock on which a kestrel is resting unseen, until it rises with reluctance as a human form approaches. From the bruised thyme underfoot a soft fragrance rises; and high in the sky above, larks are singing strongly. Much marshy ground surrounds the marl loch and sedges, cotton-grass and horse-tail abound there. The little grass-of-Parnassus is in flower here at this time. So back to the bridge, after a too brief visit to the first local nature reserve established in Britain.

SALT-MARSHES

Salt-marshes occur throughout Scotland behind the muddy shores of the firths, at the head of the western sea-lochs and in a few places on the mud trapped on the landward side of shingle spits. The conditions required for their development are the stabilising of silt or muddy sand: this process is aided by the growth of fine algal mats and the building up of organic debris in the form of decaying seaweed. Along the many creeks of the Solway, extensive salt-marshes, known locally as merselands, have been formed. Two plants are dominant here, as indeed they are in many other places in Scotland. One is the sea poa-grass *Puccinellia maritima*, a strong-growing plant which gives the merse its characteristically tussocky appearance; the other is glasswort *Salicornia* spp, a very variable plant which colonises the mud at the seaward edge of the marshes. The plants which follow this early colonisation are sea-blite *Suaeda maritima*, thrift, sea-aster *Aster tripolium*, sea-plantain, sea arrow-grass *Triglochin maritima* and scurvy-grass. The flora of Scottish marshes is rather more limited than that of English localities. Sea-lavender *Limonium vulgare* is distinctly uncommon but can be found in the Cree estuary, and at Rough Firth sea-purslane *Halimione portulacoides* grows in its typical habitat along the edges of the creeks. The cord-grass *Spartina townsendii*, which is so abundant on the mud-flats of southern England, is rare in Scotland; but stretches of it occur in Kirkcudbrightshire at the Fleet estuary and in Auchencairn Bay and it also grows along the Berwickshire coast. The basic composition of the marshes of the east-coast firths is similar to that of the Solway merselands and includes the following species not mentioned so far: creeping fescue *Festuca rubra* which is often abundant, mud-rush *Juncus gerardii* in the higher parts and sea club-rush *Scirpus maritimus*.

At the head of the western sea-lochs there is often a zone of marsh where again the sea poa-grass is abundant; but glasswort, which is normally a co-dominant, is much less common. The seaward fringes of these marshes are characterised by various species of wrack which exist in special forms found only in this habitat. These include knotted wrack *Ascophyllum nodosum* and the small *Fucus ceranoides* which lacks a common name. A member of the sedge family which resembles the club-rushes is the red blysmus *Blysmus rufus*, a plant of northern distribution which is mainly confined to the western and northern coastlines in the upper zones of the marshes. The common reed *Phragmites communis*, which grows in a few places on the Solway shore, also forms swamps on the landward side of some of these west-coast marshes where only moderately brackish conditions persist. Visitors to the west coast in June cannot fail to notice the yellow flowers of flag iris *Iris pseudocorus* which is very prevalent wherever there is an adequate quantity of fresh water to dilute the salinity.

When we come to the invertebrate fauna we find that it is a relatively impoverished one. Although the dog's-tooth *Lacanobia suasa* has been recorded from south-west Scotland, the moths are largely absent and the grasshopper and bush-cricket of the saltings do not reach this far north. The same applies to a number of plant-bugs; but a few are found. One which does occur, feeding on sea club-rush, is *Teratocoris antennatus*. Two more belong to the genus *Conostethus*: *C. frisicus* has been recorded from the Solway Firth and *C. brevis* in Britain is known only from the salt-marshes of the Scottish east coast. A few ground-beetles are particularly associated with salt-marshes and certain minute water-beetles, including *Ochthebius* spp. A tiny ladybird-beetle with bright yellow legs, *Scymnus testaceus*, is of local occurrence under plant debris. A number of spiders live on these marshes although hardly any are restricted to a saline habitat. Nearly all are dark-coloured and several are very minute, including members of the genus *Silometopus* and *Dictyna pusilla*. The last-named is a northern species of a number of inland localities but it is recorded also from the Morayshire coast. Other spiders from the north-east saltings are *Singa pygmaea*, *Arctosa leopardus* and *Lycosa purbeckensis*, a wolf-spider which is locally abundant.

COASTAL WATERS

The coastal waters are a specialised sphere belonging more to the skin-diver and to the marine biologist with his elaborate equipment than to the ordinary nature-lover standing on the shore. But he will find shells of deep-water molluscs cast up on the beach; he may find a dead fish or mammal; and he will certainly see the offshore birds resting in rafts on the sea or gliding over the waves. So for the sake of completeness a brief account seems appropriate.

A great wealth of shell life lives on the sandy and mud bottom, far too numerous to describe in any detail. Most of the scallops are free swimmers and the great scallop *Pecten maximus* with a shell nearly 6in in diameter, is very common. The seven-rayed scallop *Chlamys septenradiata* is a northern species, although it is not very common in Scottish waters. It occurs mainly from Argyllshire northwards. The queen scallop *C. opercularis* is fished commercially in the Clyde and off the east coast. There is commercial fishing of oysters too; and at several places, experimental beds have been established by the Scottish Marine Biological Association. Smaller bivalves include the common basket-shell *Corbula gibba*, the trough-shells *Spisula* spp and the dog-cockle *Glycymeris glycymeris*. The spindle-shell *Neptunea antiqua*, also known sometimes as buckie, is a large gastropod some 4in high. Of northern distribution, it occurs all round the Scottish coasts. A much smaller gastropod is the necklace-shell *Natica alderi*. Quite different from the usual pattern of molluscs, because of its curved shape, is the tusk-shell *Dentalium entalis* and this is more common off Scottish shores than English ones. The lobster *Homarus vulgaris* is a crustacean which is fished extensively, particularly by the crofters of the west coast. Nearly a million lobsters are caught annually. Landings of the Norway lobster *Nephrops norvegicus* have increased considerably in recent years and much exceed those of the common species. In 1971 the value of Norway lobsters caught amounted to £2,114,667 as against £904,899 for the common lobster, but individually the latter are much more valuable. Crabs, prawns and shrimps are other crustaceans of some commercial importance and Mallaig is an important fishing port for them.

It is always of absorbing interest to watch the fish being unloaded, particularly in the smaller west-coast harbours such as Mallaig and Kinlochbervie where, despite the bustle of activity, there is a more intimate atmosphere than is possible in the large commercial docks of the east coast. The onlooker feels to be almost part of the scene rather than an intruder, but intruders there certainly are in the shape of herring gulls, which make concerted and often successful onslaughts on the ready-served meal. Patterns of fishing change over the years due to vacillations in demand and to availability of the fish; and the most conspicuous change in Scotland has been the decline in the herring fisheries. In the eighteenth century herrings were coming into the west-coast lochs in vast numbers and Loch Broom in Ross-shire was a great centre of the industry; but a hundred years later the fishing had virtually come to an end because the seasonal movement of the herring had changed and the shoals had ceased to come into the lochs. It seems likely that this change of behaviour may well have been due to changes in the distribution and relative abundance of the larger zooplankton on which the herring feed. Loch Fyne has been and is still well known for the quality of its herrings which locally are called 'Glasgow magistrates'. Herring shoals still abound off the west coast, but in recent years from Lerwick in the Shetlands southwards there has been a shift of emphasis and whitefish have come more and more into the picture. On the west coast, Ullapool and Mallaig are the principal herring ports although the latter also handles whitefish and crustaceans. On the east coast, Aberdeen is by far and away the most important fishing port with Peterhead coming second. Offshore fish include various flat-fish, whiting, skate, cod in relatively small numbers and mackerel which move into coastal waters during the summer.

An excellent practice has been established: fishermen send in unusual fish to the Government fishery officers for identification. In this way a valuable series of records has been accumulated of many very rare fish in Scottish waters. Some have quaint names which excite curiosity: angel-fish, moon-fish, sun-fish, trumpet-fish and electric ray. That bizarre fish, the seahorse, is extremely rare around the British Isles but one was found on the Argyllshire coast at Duror on Loch Linnhe in August 1957. Shark-fishing is developing as a sport and several species are hunted. The tope is a northern species found all round the Scottish coasts and the basking shark is a summer visitor which in some years is relatively common. At times it is seen close inshore along the west coast.

From various vantage points such as Fife Ness, Tarbat Ness, Faraid Head and many others, the autumn movements of shearwaters, skuas, auks, terns and other sea-birds can be observed. On the west coast manx shearwaters are then dispersing from their breeding haunts to more southerly waters. The sooty shearwater is uncommon but fairly regular in autumn; and the great shearwater is seen during the summer as well, mainly in the Minch. In contrast to the shearwaters' preference for the west coast, the migrating skuas are more numerous on the east coast and of the four species the long-tailed skua is the one least often seen. Little gulls are increasing in numbers on the Fife coast and can often be seen in summer also.

Winter finds numbers of divers, grebes and sea-duck on the coastal seas; and as well as the resident gulls there are occasional glaucous and, less often, Iceland gulls. All three divers are common and non-breeding great northern divers sometimes remain during the summer in the north-west. Numbers of great-crested grebes move to the coast in winter: they frequent certain localities more especially on the east coast, such as at St Andrew's Bay in Fifeshire. Black-necked, red-necked, Slavonian and little grebes also inhabit coastal waters, but to a much lesser extent. Cormorants and shags are generally distributed and both sometimes gather in flocks in winter. Of the ducks, mallard, tufted, red-breasted merganser and eider are generally common on both west and east coasts; but some ducks are more plentiful on one side of Scotland than the other. Thus, species such as scaup, golden-eye and velvet scoter are rare or absent in the north-west although present in considerable numbers in some other localities. On the other hand, long-tailed duck are scarce in the south-west but increase northwards, becoming common on the north-west, north and east coasts. Common scoters, though widespread, are more numerous on the east coast than on the west and flocks are sometimes seen even in summer. Occasionally much rarer duck, such as king eider and surf scoter, are reported, although these may not be quite as rare as is usually thought. They are normally seen from autumn to spring; but in June 1971 a drake surf scoter was present in a mixed flock of common and velvet scoter off the Aberdeenshire coast.

Turtles, or sea-tortoises, are strange, plated reptiles which are occasionally observed in Scottish waters and infrequently found washed up on the shore. The lath is a large species, 8ft in length: a fine specimen can be seen in the Royal Scottish Museum in Edinburgh, and one was found dead on the inner Solway shore in August 1959.

Both common and Atlantic grey seals are frequent round the coastline. The latter, as its name implies, is much more a mammal of the west coast, especially of the more remote skerries and islands where it breeds; but it occurs on the east coast and visits firths which are the haunt of the common seal. The latter is widespread but this too is more plentiful on the west. The grey seal is a distinctly larger animal and the shape of its head is different, with a dissimilar nostril pattern, but these features are often impossible to distinguish with certainty when all the observer has is a momentary glimpse of a bobbing head among the waves. The elegiac crying of the seals may sound weird to some ears but it is perfectly in keeping with the grey wastes of the ocean. It forms a theme in a fugal composition in which the roar of the Atlantic breakers and the mewing of the gulls are other components. The two species have quite different breeding seasons: pups of the common seal are born in midsummer but those of the grey seal in the autumn. Both species are unpopular with fishermen for alleged damage to salmon fisheries and licences to shoot seals are issued under the 1970 Conservation of Seals Act. The walrus is a mammal of the far north-western Atlantic and a relative of the seals. It used to be seen with some regularity in Scottish waters, but in this century its range has contracted and none had been observed between 1928 and 1954 until, in February of the latter year, a live one was seen off the Aberdeenshire coast.

If the observer is often unable to obtain as good a view of a seal as he would like, how much more is this the case where whales are concerned. To the land-lubber they are all very similar; but the occasional glimpse of a fin or a body surface is inadequate for identification purposes. Yet during the summer months there is a considerable migration of a number of species in northern waters; and it is often possible for the traveller on one of the west-coast steamers in the Minch to have a sighting of one of these marine mammals. Strandings occur occasionally, and it is interesting to note that the law of the land regarding stranded Royal Fish in Scotland differs from that in the rest of Britain. In this connotation Royal Fish includes mammals of the order Cetacea. Whales, porpoises and dolphins generally belong to the Crown; but in Scotland those that are less than 25ft in length together with bottle-nosed and caa'ing-whales of any size, are not Royal Fish and are excluded from the provisions of the Act.

The large whalebone whales usually keep well away from the coasts but the common and lesser rorquals are fairly frequent in the Minch. This latter species can sometimes be seen close inshore in late summer all round the coasts of northern Scotland. The bottle-nosed whale is fairly common on its southerly migration in the autumn and there are regular sightings of it in the Firth of Clyde. The dolphins are a family of toothed whales, on the whole considerably smaller than the whalebone whales. The white-beaked dolphin is mainly concentrated on the eastern coast, although it is reported commonly in the Firth of Clyde. The caa'ing or pilot-whale is a northern species of dolphin which used to be hunted in the northern isles when shoals came close inshore. The large and ferocious killer whale, which hunts in small groups, is also a dolphin. The commonest cetacean in Scottish coastal waters must be the porpoise. In contrast to the dolphins, which generally steer well clear of boats, a school of porpoise will appear to show an interest in a boat by surfacing near it and swimming around it.

CHAPTER FIVE

Offshore islands

The Clyde islands–Islands of the north-west–The Forth islands

SMALL ISLANDS HAVE always had a great attraction for the naturalist: perhaps it is an atavistic urge going back before the concept of an Englishman's castle and stemming from the Nordic rover part of our ancestry. It has to be remembered, however, that islands have an impoverished fauna compared with the mainland, although in the case of islands long separated from the adjacent coast this may be partially offset by the existence of newly evolved subspecies.

There are a few islands in the Solway Firth, but little need be said about these as most are small areas of land joined to the mainland at low tide. In the centre of the wide entrance to Luce Bay in the outer Solway are the Scar Rocks. The largest of these stacks has a gannetry which is increasing in size. A colony may have existed at an earlier date but breeding was not satisfactorily proved until the beginning of World War II; by 1971 there were 500 nests.

THE CLYDE ISLANDS

Rounding the Mull of Galloway and moving north we come to the approaches to the Firth of Clyde, where 10 miles due west of Girvan is the small but well-known island of Ailsa Craig. It is seen from the Ayrshire coast as a conspicuous broad-based cone, the peak of which in summer is white with gannets. The island, with very steep cliffs, is nearly a mile in length and rises to 1,110ft in height. It is composed of microgranite, an acid

Page 89 (*above*) Red-throated diver, incubating; (*below*) the Black Wood of Rannoch, a remnant of the old Caledonian forest which supports a rich flora and fauna

Page 90 (*above*) Pedunculate oak wood in the Dee valley with clump of great wood-rush in foreground; (*below*) sika stag at Dawyck, Peeblesshire

igneous rock with relatively fine-grained crystals of quartz, felspar and mica. It is actively quarried.

The cliff plants are thrift *Armeria maritima*, scurvy-grass *Cochlearia officinalis*, tree-mallow *Lavatera arborea* and red campion *Silene dioica* which make a colourful assemblage in the summer months. Sea-radish *Rhaphanus maritimus*, which in Scotland is a rare plant, grows on the ledges; and a common plant which grows on the slopes as well as the ledges is wall pennywort *Umbilicus rupestris*, an indicator of the acid soil derived from the underlying granite. It has been noted by H. G. Vevers that, where rich guano deposits have accumulated on some of the ledges, thriving communities of the common stinging nettle *Urtica dioica*, annual poa-grass *Poa annua* and red campion have resulted. There are screes composed of granitic boulders on which wood-sage *Teucrium scorodonia*, ragwort *Senecio jacobaea*, bracken *Pteridium aquilinum*, hairy bitter-cress *Cardamine hirsuta* and heath bedstraw *Galium saxatile* grow in quantity. The welted thistle *Carduus acanthoides* is also frequent here although generally rare elsewhere in Scotland.

The common butterflies of grassland can be seen flying in the sunshine, but much smaller numbers are recorded of large heath *Coenonympha tullia*, grayling *Hipparchia semele*, brown argus *Aricia agestis* and small copper *Lycaena phlaeas*. The gannetry, which is situated chiefly on the southern side, is probably one of the largest in the world and continues to increase; a census in 1951 revealed 7,833 pairs, whilst the count in 1971 enumerated 14,347 pairs which by any standards is a notable increase. It is a long-established colony known to have been in existence in 1526. The breeding population of common guillemots has been reduced in recent years and there are now about 4,000 pairs. Razorbills and puffins nest in much smaller numbers: the population of the former has kept fairly constant, but that of the latter has been subject to marked periodic fluctuation, with a general downward trend, from the vast numbers of the last century to a few pairs at the present time. Other sea-birds are fulmars, a few pairs of cormorants and the cliff-nesting seagulls including a large colony of nearly 8,000 pairs of kittiwakes. Black guillemots only began nesting about nine years ago, but a small colony is now established. Slow-worms and common lizards are fairly plentiful and there are small numbers of palmate newts. Rabbits are numerous and among the small mammals pygmy shrews are found. There used to be a small herd of wild goats but this has died out in the last five years. (It will be interesting to see whether the larger plants such as the tree-mallow and the elder bushes will increase now that the goats have gone.) A similar disaster seems to have overtaken the Soay sheep about the same period. Common seals are usually the species seen in the sea around Ailsa.

Three miles south-west of Troon on the Ayrshire coast is Lady Isle. There are tern colonies nesting here, principally common, arctic and sandwich terns. The reserve used to have a large number of roseate terns, but this species is notorious for its fluctuations in population and only two pairs nested in 1971. Horse Island is a similarly low-lying but rather smaller island a mile offshore of Ardrossan. It has a large colony of black-headed gulls and four other species of gull nest here. Four species of tern breed, but the roseate population has declined here as well.

Little Cumbrae is a privately owned rocky island in the mouth of the Clyde, 1 mile long by ½ mile wide. Its cliffs are frequented in summer by a pair of ravens, several pairs of black guillemots and numbers of gulls. There are tern colonies and as well as the usual waders there are a few pairs of woodcock. Eider, red-breasted merganser, shelduck, mallard and teal breed here. Although the island is not large it has a number of land birds including buzzard, red grouse, whinchat, stonechat, wheatear, twite, linnet and redpoll. Of special note are the pair or so of nightjars, as this is a distinctly local species in Scotland.

About ½ mile farther north is the larger island of Great Cumbrae. The rocks consist principally of Old Red Sandstone, with some igneous dykes. Agricultural land alternates with areas of moor and small belts of mixed deciduous woodland. The coastline is varied with sandy beaches, many rocky stretches, dunes and salt-marsh. There is a raised beach on which grow moisture-loving plants such as butterwort *Pinguicula vulgaris*, sedges *Carex* spp, rushes *Juncus* spp and flag *Iris pseudocorus*. A wealth of marine life is found along the coast and until recently the Scottish Marine Biological Association was based at Millport in the south of the island. It has now moved to Oban and the marine station at Millport is operated by Glasgow and London Universities. Millport Bay is of historic interest in that the roseate tern was first identified as a separate species here by Dr MacDougall in 1812 and the scientific name gives recognition to the discoverer. The fieldmouse on Great Cumbrae is a form of the Hebridean fieldmouse which has been given subspecific rank *Apodemus hebridensis cumbrae*.

The island of Bute to the north-west is much larger than the Cumbraes combined, being some 16 miles in length and 4 miles in width at its broadest part. There are two well-marked constrictions in the outline which give the island the shape of a giant insect: the narrow head projects out into the centre of the Clyde, and there is an elongated thorax behind the neck at Kilchattan, followed by a truncated abdomen behind the wasp-waist at Kames Bay. The northern part and the extreme south are hilly, but most of the island consists of green pastures and fertile arable land. Geologically, it is cut in two by the diagonal bisection of the Highland Boundary Fault: to the south is Old Red Sandstone with some basaltic lavas and to the north is the much older Dalradian Schistose Grit.

The mixed deciduous woodland contains sycamore, beech, ash, oak and birch; and there are larch and spruce plantations. These woods have the common ferns of hilly northern districts such as the oak-fern *Thelypteris dryopteris* and beech-fern *Thelypteris phegopteris*, as well as the more local, hay-scented buckler *Dryopteris aemula*. Interesting coastal plants are the sea-spleenwort *Asplenium marinum*, oyster-plant *Mertensia maritima*, lovage *Ligustum scoticum*, Danish scurvy-grass *Cochlearia danica* and brookweed *Samolus valerandi*. Inland waters have the yellow water-lily *Nuphar lutea*, the white water-lily *Nymphaea alba*, the alpine pondweed *Potamogeton alpinus* and the floating marshwort *Apium inundatum*.

Since much of the land is agricultural and the coastline is largely low-lying, the bird life is not outstanding. There are no snakes on Bute, but common lizards, slow-worms, smooth newts and toads are well distributed whilst frogs and palmate newts are even more plentiful. The only subspecies of the mainland fieldmouse *Apodemus sylvaticus* lives on Bute and is named *A. s. butei*; it differs from the type in being darker in colouration, with a rather shorter tail and ears. It is an indication that, although the island appears to be almost a part of the mainland, it has in fact been separated for a long enough time for an island race to evolve. The other common small mammals occur, but water-shrews are scarce. Farmland mammals such as moles, hedgehogs, stoats and weasels are plentiful, but no badgers or foxes are known. The woods are inhabited by roe deer and the hilly parts of the island have had two small herds of wild goats. The herd on the northern hills is still in existence; but there are conflicting reports on the one in the south which may now be extinct. G. Kenneth Whitehead states that the best goat-head in Scotland, with a horn-length of 44¾in, came from Bute.

South-west of Bute lies the island of Arran in the lea of Kintyre. It is 20 miles long by 10 miles at its widest part; it has a 60-mile coastline and an area of 165 square miles. Arran has often been described as a microcosm of the Highlands: this is a fair description, for the island stands in marked contrast to the neighbouring Bute. Although Windy Hill on Bute reaches to 911ft, Goatfell on Arran attains 2,866ft and nearly half the island is over 1,000ft above sea-level. The northern part of the island in particular is of high scenic value, with picturesque glens and hanging valleys. The lower hill slopes, river valleys and coastal land are farmed; but much of the higher land is kept as deer forest. The coastline is varied with towering cliffs, caves, sandy bays, dunes and rocky beaches with rock-pools. Raised beaches, mainly at the 25ft level, virtually encircle the island.

Arran is of great geological interest, with varied rock types. The area around Goatfell is an igneous complex consisting principally of granite intruded into surrounding sedimentary rocks of New and Old Red Sandstone and metamorphic rocks of Dalradian Schist. This intrusion of granitic magma probably took place in Tertiary times and resulted in an uptilting of the older existing rocks, so that today a ring of upturned strata encircles the granite. There are numerous volcanic dykes which stand out clearly because the surrounding sandstone is more susceptible to erosion, but in some instances, where the dykes bisect granite, the opposite process has taken place. The glaciers of the Ice Age have deposited many erratics from distant places. In the north-east there are Coal Measures and coal was at one time mined in small quantities.

On the rock ledges and slopes above the glens grow typical mountain plants like starry and purple saxifrage *Saxifraga stellaria* and *S. oppositifolia*, cloudberry *Rubus chamaemosus* and moss-campion *Silene acaulis*. The rare alpine lady's-mantle *Alchemilla conjuncta* grows in Glen Sannox at one of only two British stations.

The moors have ling *Calluna vulgaris* as their basic constituent, but Arran has a high rainfall and there are many boggy areas with cross-leaved heath *Erica tetralix*, purple moor-grass *Molinia caerulescens* and butterwort *Pinguicula* spp. The few-flowered sedge *Carex pauciflora* is a small and uncommon sedge virtually restricted to northern moors. It was first identified as a British plant by the Rev John Lightfoot in 1772 when he discovered plants growing in Arran.

Among the numerous moths of the moors may be mentioned the Arran carpet *Chloroclysta concinnata*, which in Scotland is confined mainly to this island. It bears a marked resemblance to the common marbled carpet *C. truncata* but the habitats and breeding pattern are different. An arctic species which reaches its southern limit in Arran is the northern arches moth *Apamea exulis assimilis*. The Scotch Argus butterfly *Erebia aethiops* was first identified by Sir Patrick Walker in 1804 on moorland near

Brodick where it still occurs. At the same time he captured specimens of a European butterfly *E. ligea* not known before in Britain. The vernacular name then given to it was Arran brown, but there is only one other record for this butterfly in Scotland and repeated efforts have failed to locate it again.

All the characteristic birds of moorland can be found even if some are scarce. Red grouse are plentiful but black grouse may be approaching extinction. Arran lies in the zone of overlap of carrion and hooded crows and both species occur. Dunlin, golden plover and curlew nest here, but only the last-named is common. There are several species of predators though none is numerous; they include golden eagle, buzzard, hen-harrier, peregrine, kestrel, merlin and short-eared owl. A large population of red deer frequents the moors and a herd of wild goats inhabits Holy Isle in Lamlash Bay. Some mountain hares have recently been introduced in one locality and may become established.

The glens are well wooded with mixed deciduous trees of oak, ash, birch and rowan. Two rare whitebeams *Sorbus pseudofennica* and *S. arranensis* are endemic to Arran. The common woodland flowers abound and the alpine enchanter's nightshade *Circaea alpina* is very local in deep shade in its only Scottish station here.

The small mammals are well represented and bank voles are reported to be increasing in numbers. The fieldmice on Arran are a separate race of the Hebridean species and have been named *Apodemus hebridensis fiolagan*. Red squirrels are widely distributed but the badger population has declined almost to vanishing point and there are no roe deer.

The farmland has pheasants and partridges although these are not plentiful. Some of the rabbits have survived myxomatosis and brown hares are relatively numerous, but foxes are absent. About forty years ago some hedgehogs were introduced and these have gradually increased and spread over most of the island.

Dippers, kingfishers, pied and grey wagtails haunt the streams.

There are colonies of common gulls on the hill-lochs and in recent years small numbers of black-headed gulls have begun to nest. The red-throated divers nesting on the lochans are at the southern limit of their range. Otters are fairly common and mink have arrived. All three species of newts live on the island but the palmate newts are the most numerous.

The shores provide much of interest to the naturalist. The oyster-plant grows in the south-west and the local carline thistle on the rocks near Bennan Head. There is a rich marine life and the cliffs have sea-bird colonies. The seals seen are usually the common species which breeds in several localities on the island.

Two miles off the southern tip of Kintyre lies a small group of tiny islands and skerries of which Sanda is the largest, rising to 405ft at its highest point. Sheep Island and the even smaller island of Glunimore have for long been the most important breeding haunts of the puffin in the Clyde area. Shags are numerous on the three largest islands and small numbers of black guillemots breed. In contrast to the other Clyde islands the Atlantic grey seal is the one most frequently seen here.

ISLANDS OF THE NORTH-WEST

The islands off the west coast from Kintyre to the Applecross peninsula are part of the Inner Hebrides and are therefore outside the scope of this volume. Beyond Applecross the first group of islands is that of the Summer Isles in the entrance to Loch Broom. There are over twenty-five islands plus numerous skerries spread out over some 30 square miles of sea and they can be visited by boat from Ullapool and Achiltibuie. On a fine summer's day it is a delightful experience to cruise among these idyllic rocky islands and reefs of Torridonian Sandstone—purple gems set in a malachite sea—while fulmars, gulls and terns fly overhead.

The largest island of the group is Tanera More of 800 acres with several small lochs and a hill 406ft high from which, on clear days, the Outer Hebrides can be seen. A larger number of breeding birds have been recorded here than on any of the other islands. There is much dense heather and bracken, and bluebells were in flower near the pier on the day I stepped ashore.

Horse Island lies nearer Ullapool, close to the northern shore of the outer approaches to Loch Broom. Its heather-clad rocks are the home of a herd of wild goats which were first noted in 1937 but which, it is believed, had been there for some years before that. One and a half miles south is the rocky islet of Carn nan Sgeir: actually two islands linked by a causeway of shingle on which terns sometimes nest. The low cliffs are heavily covered with orange, grey and black lichens and the cliff-tops have a grass turf kept short by rabbits and sheep. The rock ledges are studded with thrift, sea-campion, scurvy-grass and dense masses of chickweed. There is a strong society of the great wood-rush *Luzula sylvatica* and here and there are plants of lovage. Herring gulls and shags nest on the cliffs, oyster-catchers and rock-pipits along the shoreline.

West of Tanera More are dozens of islets and skerries, the kind of environment where one would expect to see seals in plenty; certainly they occur occasionally for I have seen them myself, and they have bred on Tanera Beg, but they are not abundant. Tanera Beg is one of the few places in the north-west which has a beach of coral sand. The two outermost islands are Glas-leac Beag and Eilean a' Chleirich or Priest Island. The first-named is a grassy island of 34 acres inhabited by greater black-backed gulls. Greylag geese use the island as their moulting ground in high summer and a large flock of barnacle geese winter on the island. Priest Island is much larger, being 300 acres; it is difficult of access due to heavy ground swell and the absence of a proper landing-place. It seems incredible that a human population could ever have lived in such a small and remote spot, but a few crofters once maintained a precarious existence there. None of the Summer Isles is inhabited now, Tanera More being the last to be occupied. Priest Island has many fewer breeding species of birds than has Tanera More but storm-petrels occupy the rock crannies, cormorants as well as shags sometimes nest and the island once held a peregrine eyrie.

Farther north beyond the Point of Stoer, in the fine sweep of Eddrachillis Bay, a similar group of rather smaller islands lies at the entrance to Badcall Bay. They are more difficult of access but have many birds including cormorants. Five miles north of Badcall is the larger island of Handa, 766 acres in size and separated by a narrow sound from the mainland. Handa is an isolated mass of Torridonian Sandstone lying off a foreland of Lewisian Gneiss and it has been renowned for centuries for its wealth of bird life. On its north-west side stands an impressive, detached rock known as the Great Stack, whose 400ft high cliffs teem with thousands of nesting sea-birds.

Torridonian Sandstone does not usually support a rich flora, so that Handa with its 216 species is fortunate in this respect. The cliff ledges have the usual maritime plants and primroses, bluebells, red campion and thyme grow among the grass on the cliff plateau, where there are also masses of crowberry *Empetrum nigrum*. Two ferns growing in crevices are the sea-spleenwort *Asplenium marinum* and the black spleenwort *A. adiantum-nigrum*. Growing on the east coast is hybrid marram *Ammocalamagrostis baltica*, the result of crossing between marram and bush-grass; it is known only in two other areas in Britain, Northumberland and Norfolk. The burns are flanked with flag iris and in one locality there is a clump of royal fern *Osmunda regalis*. Near the mouth of the little burns grow plants of the maritime form of the very local water whorl-grass *Catabrosa aquatica* var *littoralis*. There are several freshwater lochs with water-lily *Nymphaea alba occidentalis*, bog pondweed *Potamogeton polygonifolius* and broad-leaved pondweed *P. natans*. The centre of the island consists of moorland with much deer-grass *Scirpus caespitosa* and purple moor-grass *Molinia caerulescens*; cotton-grass *Eriophorum* sp grows in the bogs; eared willow-scrub *Salix aurita* is found in sheltered moist places; and bracken, ling and bell-heather grow in the drier parts. For a small island there is a rich moss flora with over a hundred species. There is a varied marine life in the rock-pools with periwinkles *Littorina* spp, beadlet anemones *Actinia equina*, jellyfish *Aurelia aurita* and various small fishes. Ground-beetles of a number of species are plentiful and, somewhat unexpectedly, I have seen on the beach numbers of the 'woolly bear' caterpillars of the garden tiger-moth.

But above everything else Handa is a bird island. The cliffs hold colonies of the order of 30,000 pairs of guillemots, 6,000 pairs of razorbills, 7,000 pairs of kittiwakes and 2,000 pairs of fulmars, with lesser numbers of puffins, shags, herring gulls and great black-backed gulls. The shingle beaches have ringed plovers, oyster-catchers and rock-pipits. The first-named is not content to rely upon the cryptic colouration of its plumage but frequently indulges in distraction displays; and I have seen one blown over by the wind on a Handa beach whilst it was engaged in such a display. A speciality in the bird life of Handa for many years was an albino oyster-catcher which had a deformed, down-curved bill. This eccentric character had reached the ripe old age of nineteen when it died on 9 July 1967; it is now preserved in the Royal Scottish Museum, Edinburgh. Offshore during the summer, shelduck and eider ducklings can be seen in stormy weather, bouncing up and down like corks among the wind-lashed waves, whilst hovering menacingly overhead will always be herring gulls and great black-backed gulls awaiting their opportunity. A pair of red-throated divers sometimes nest on one of the lochs and in 1964 the first pair of great skuas bred on the island. Wheatears and twites have territories in the moorland areas and merlins have attempted to nest. Handa had for many years an eyrie of the white-tailed sea-eagle, but the last pair ceased to nest about 1864. Brown rats were introduced about 1870 and they, together with rabbits, are numerous.

THE FORTH ISLANDS

There are few islands on the east coast except those in the Firth of Forth. Of these the largest is the Isle of May, which is situated 6 miles south-east of Fife Ness at the entrance to the firth. The site of a long-established bird observatory, it is 140 acres in size, and from 1956 has been a National Nature Reserve. Access is usually by boat from the fishing villages of Anstruther and Pittenweem in Fife, on the north shore of the Firth of Forth. Of all the islands looked at so far the Isle of May is the one whose plant and animal life have been the most intensively studied over a long period, in fact from 1907 onwards. It is possible here to give only a brief account of the principal communities. Those seeking a fuller account are referred to the comprehensive book by Dr W. J. Eggeling.

From 150ft high cliffs on the west side the surface slopes to sea-level on the east coast. Several west-east faults occur; three of these fractures have been eroded by the sea, causing the island to be divided into four segments. There is only one main rock type, a dark grey igneous rock in the form of a sill of molten matter intruded horizontally between sedimentary strata. Only the igneous material now remains and this is a basic rock of intermediate type known as dolerite: it contains the green mineral olivine.

It is interesting to compare the Isle of May vegetation with that of a west-coast island such as Handa. The principal differences lie in the absence on the May of bog and heath communities and the presence of many waste-ground plants and garden weeds: the former are due to the extensive ash deposits from the old coal-beacon which served as a lighthouse from 1636 to 1816 and the latter to the cultivations by the lighthouse keepers. As on Ailsa Craig there is a rich plant community resulting from gull guano in nesting and roosting colonies on the two northern islets, although the constituents differ from those on Ailsa in that they consist principally of sorrel, chickweed, cleavers *Galium aparine* and curled dock *Rumex crispus*. As one would expect, lovage grows in some abundance on ledges and on slopes. The fern, sea-spleenwort *Asplenium marinum* grows on the cliff, and there is a strong colony of the uncommon sea-wormwood *Artemisia maritima* in the south of the island. A number of the plants must have arrived by seed in bird-droppings and a bizarre example of this is the fig-tree, recorded by Dr Eggeling, which grows in a cleft on the east coast.

Offshore islands are not well endowed with butterflies and the May is no exception. Only six species are regularly seen: the whites *Pieris* spp, small tortoiseshell *Aglais urticae*, painted lady *Cynthia cardui* and red admiral *Vanessa atalanta*. A considerable number of moths have been recorded, including the garden tiger-moth, the caterpillars of which species, as noted earlier, were observed on a Handa beach. The northern rustic *Standfussiana lucernea* is a moth whose habitats include rocky coastal areas; it has a discontinuous distribution, occurring in southern England and then from Yorkshire northwards. The square-spot dart *Euxoa obelisca* and the marbled coronet *Hadena confusa* are other coastal species. Many other common moths have been seen and the square-spot rustic *Xestia xanthographa* is especially plentiful. The various families of flies are well represented, but the exposed coastal situation means a welcome reduction in the biting flies: only one species of biting midge and two of Tabanids have been recorded.

The western cliffs have strong breeding colonies of sea-birds. There are puffins as well as guillemots and razorbills and it is pleasing to record that although for more than the first half of this century the puffin population on the island did not exceed 50 pairs, there has been a rapid increase in the last few years up to a total of 2,500 pairs in 1971. The shag is another bird which has made a dramatic increase from the 10 pairs recorded by Dr H. N. Southern in 1936 to several hundred pairs today. Fulmars have bred from 1930 onwards but are not very plentiful. There was a population of over 3,400 pairs of kittiwakes in 1971 and there are small numbers of lesser black-backed gulls. Herring gulls have become so numerous that it has been necessary to institute control measures. Starlings, rock-pipits and oyster-catchers are other breeding birds around the coastline whilst eiders, meadow-pipits and wheatears nest on the plateau. It is, however, as a migration watch point and ringing station that the Isle of May is chiefly known. It lies on the route of north-south migrants and, when weather conditions are right, it receives drift migrants from Europe. Many rarities have been recorded from time to time although the May cannot compete in this respect with its more famous rival, Fair Isle. Only two mammals, the rabbit and the house mouse, have been recorded as residents, although numbers of grey seals frequent the offshore rocks and common seals are occasionally seen.

The small islands of the Firth of Forth can be split into two groups: firstly there are those which are clustered in the approaches to the Forth Bridges, that is, Inchkeith, Inchcolm and Inchmickery; secondly there is a group of even smaller islands close inshore between Gullane and North Berwick—Eyebroughty, Fidra, Lamb and Craigleith. Some of these islands are of interest to the naturalist for their colonies of sea-birds, principally terns. In the first group, Inchmickery is a reserve of the Royal Society for the Protection of Birds. Three acres in size it held the largest nesting colony of roseate terns in the area, but in 1971 hardly one of this species was seen on the island. This may be due to one of the periodic shifts of site to which this bird is particularly prone, or partly, perhaps, to the increase in the herring gull population. Inchcolm is a larger island nearer the north shore of the firth and boat-trips to it are sometimes made in summer from Granton Harbour; a few pairs of roseate terns bred there in 1971. The largest island is Inchkeith, whose breeding birds include a growing colony of kittiwakes: this colony increased by 19 per cent to 407 pairs in 1971. Three islands of the second group are bird reserves consisting of volcanic rocks of the Carboniferous era. Eyebroughty is of interest in that it is a moulting ground for a large number of eider duck in summer. Fidra has strong colonies of terns: common and sandwich terns are the most numerous, with smaller numbers of roseate and arctic terns. Fulmars and kittiwakes breed and the first nesting pair of shags was recorded in 1971. The last and smallest of the reserves is that of the Lamb whose cliffs hold numbers of common guillemots, kittiwakes, shags and at 280 pairs the largest colony of cormorants in the Forth area. Craigleith lies nearly a mile out at sea from North Berwick and its steep cliffs have many sea-birds.

Farther out, just over 3 miles from North Berwick is the rather larger and more famous island of the Bass Rock, 313ft high and a mile in circumference. This consists of a fine-grained, acid volcanic rock known as phonolitic trachyte: the adjective describes the bell-like sound emitted when the rock is struck with a hammer. The flora can be classified into cliff vegetation, plants of the grassy plateau and weeds introduced in the cultivations arising from human occupancy in the past: monastic cell, barracks, prison and lighthouse. Common scurvy-grass, sea-campion, sea lavender *Limonium vulgare*, sea beet *Beta maritima* and Babington's orache *Atriplex glabriuscula* var *babingtonii* grow along the rocky coast. In the grass on the island summit are such plants as mouse-ear hawkweed *Hieracium pilosella* and dandelion *Taraxacum officinalis* whilst alongside paths and on waste ground are clumps of stinging nettle, welted thistle, sow-thistle *Sonchus oleraceus* and chickweed. Particularly colourful are the masses of red campion beside the paths and the dark-veined purple flowers of the tree-mallow.

But the Bass Rock is pre-eminently a bird island and, more specifically, a gannet island. The breeding colony was known to be a strong one in 1447 and if the passage about gannets in the Anglo-Saxon poem *The Seafarer* relates to the Bass, then the colony has been there for over 1,200 years. There are about thirty such colonies in the world, but such is the strong association of the gannet with the Bass that from it the scientific name *Sula bassana* is derived. The population is increasing here as elsewhere and there may now be over 8,000 pairs. It is interesting to note that in the southward autumn migration the young birds travel farthest, reaching the west coast of Africa, whilst the older ones in the main prefer to frequent European waters. The island occupies approximately the same position on the southern side of the firth as the Isle of May on the northern shore, but the Bass Rock has little visible migration. Herring gulls and lesser black-backed gulls nest on the plateau and numbers of auks, fulmars and kittiwakes on the precipitous cliffs. A good vantage-point from which to see sea-birds other than gannets is the ravine on the west side of the island. There, kittiwakes with their young can be seen across the chasm, while many feet below, shags perch on stacks rising from the sea and in the foaming waters around them common seals toss and roll. The year 1967 saw a remarkable addition to the Bass Rock bird fauna, with the presence throughout the summer of a very rare visitor from the Antarctic: a black-browed albatross. These birds of southern seas are seldom seen in northern waters and for one to frequent an island so close to the mainland and for so long a time was an unheard-of event. It must have liked what it saw of Scotland for it returned in 1968; but in that year it spent much more of its time away from the Bass, and I count myself fortunate in having seen it that year on a day of thick fog. In later years this species has been observed off the northern isles and it is possible that a change in movements outside of the breeding season is beginning to take place.

CHAPTER SIX

Aquatic habitats

Rivers—Lochs

RIVERS

THE NATURE OF Scottish rivers is much influenced by the character of the underlying rocks. The main watershed is sited much nearer the west coast, with the result that the west-flowing burns have a rapid fall to the sea in contrast to the longer and larger east-flowing rivers. The hard quality of the substrate means that the smaller streams are able to gouge out their course only through a narrow valley or gorge. The larger rivers flow through wide straths and are flanked by haughlands which is the Scottish term for alluvial meadows.

The Clyde, 106 miles in length, is the longest river on the west coast. It rises in the Southern Uplands west of Moffat and flows northwards to Glasgow, forming on the way the Falls of the Clyde at Corra Linn. The many short, west-flowing mountain streams are often exceedingly attractive as they cascade over dipper-haunted rocks between banks fringed with rowan, birch, and alder. One of these, the river Elchaig in Wester Ross has one of the largest of Scottish waterfalls in the Falls of Glomach: these are separated into three stages of which the middle one, 220ft high, is by far the largest. Three small burns in the Cambrian Limestone of Sutherland follow a common habit of limestone streams in running underground for part of their course. These are the Allt-nan-Uamph and Traligill burns at Inchnadamph and the north-flowing Smoo at Durness. The Allt-nan-Uamph has a glacial relict in a springtail *Onychiarus schoetti* which lives in an underground pool.

Page 107 Cock capercaillie, a characteristic bird of the old pine forest

Page 108 (*above*) Raspberry fields in Angus which attract fruit-eating birds in August; (*below*) wood cranesbill on a Perthshire verge

Of the east-coast rivers the Findhorn, Spey and Deveron flow northwards into the Moray Firth. The Spey is noted particularly for its swift currents, although in its upper reaches it has a gradient of only 1 in 1,260 and in the parish of Insh a large river-marsh has formed. The Don and the Dee are twin east-flowing rivers of Aberdeenshire, the former flowing through pleasant undulating countryside and the latter passing through the wooded valley of Royal Deeside with its more familiar scenic attractions. Farther south, the Tay in Perthshire issues out of Loch Tay and flows 118 miles to the sea; it is Scotland's largest river with a catchment area of 2,000 square miles. South of the Tay the Forth rises in the Trossachs, bisects the giant Flanders Moss and flows in a quite exceptional series of meanders across the alluvial plain of Stirling to its outfall in the Firth of Forth. In the Southern Uplands the Tweed and its most important tributary, the Teviot, are renowned salmon rivers in a fertile basin of Old Red Sandstone. Most Scottish rivers are of high quality and pollution is not the major problem that it is elsewhere in Britain, although some of the Central Lowland rivers have a degree of contamination.

The swift current of most of the rivers means that floating vegetation is largely absent. The onlooker leaning on the parapet of a bridge will see no dense masses of water crowfoot or of other aquatics which are so typical of the slow-moving English streams; instead he will see the sparkle of the water and hear the thunderous turbulence as the waters swirl and eddy between and over the rocks. In the larger rivers, particularly in those in the Southern Uplands and Central Lowlands, there is a certain amount of submerged vegetation. Such aquatics include the Canadian pondweed *Elodea canadensis*, water milfoil *Myriophyllum* spp, willow-moss *Fontinalis pyretica* and *Rhacomitrium aciculare*, the last-named frequent on submerged boulders in the southern Grampians. There are a few slow-moving waters where there is little fall, but for various reasons even these, in the main, are devoid of surface plants. In Perthshire the Pow Water is a very sluggish tributary—twice-removed—of the Tay; it flows through a raised moss, parts of which resemble the Fenland alder carrs, but the agricultural need for improved drainage has resulted in the stream's having been cleaned out, so that even this is bereft of floating plants. The river Endrick in Stirlingshire is exceptional in having in its lower reaches some surface vegetation: broad-leaved and curled pondweeds *Potamogeton natans* and *P. crispus*, yellow water-lily *Nuphar lutea* and duckweed *Lemna gibba*.

A conspicuous flower which has become naturalised on the midstream shingle banks and gravelly margins of a number of Highland rivers is the lupin *Lupinus nootkatensis*, the blue flowers of which make a colourful display in June. Another naturalised plant found occasionally on riverside shingle is the tall Indian balsam *Impatiens glandulifera*, with its large rose-coloured flowers; and in summer some shingle islets are a blaze of yellow from the yellow mountain saxifrage *Saxifraga aizoides*. Surprising plants may turn up in this habitat and as many as forty aliens have been recorded from shingle in the lower reaches of the Spey. River banks, stream margins and moist ground beside the water have a considerable variety of plants including alpines brought down to lower levels by winter floods. An example of this is the northern rock-cress *Cardaminopsis petrae* which has been carried down from the Cairngorms to the shingle banks of the Dee. The sword-like leaves of the yellow flag *Iris pseudocorus* stand sentinel beside many a west-coast stream. The globe-flower *Trollius europaeus* is an aristocratic member of the buttercup family, growing in hilly districts on streamsides and in other moist places. It has been recorded at a height of 3,700ft; but the photograph on page 71 shows plants growing almost at sea-level on the north Sutherland coast. Very attractive and showy plants are the two monkey-flowers *Mimulus guttatus* and *M. luteus*, especially the latter with its red-blotched yellow flowers which grow in masses along the borders of some streams and brighten the surroundings on the dullest of days. Both are naturalised, as is the variable pink- or white-flowered claytonia *Montia sibirica* which grows on some shaded river-banks. Another alien is the New Zealand willow-herb *Epilobium brunnescens*, a prostrate plant which occurs along streams but thrives also in other wet habitats. This flower first arrived in the British Isles in 1908, but it was not until 1933 that it began spreading northwards in Scotland from a centre in Glasgow and it has now reached western Inverness-shire. Other flowers of the riverside are marsh-marigold *Caltha palustris*, meadowsweet *Filipendula ulmaria*, water forget-me-not *Myosotis palustris*, marsh-thistle *Cirsium palustre*, reed-grass *Phalaris arundinacea* and reed-fescue *Festuca altissima*, the last-named a grass mainly of northern distribution.

The orange-tip butterfly whose larvae feed principally on the cuckoo-flower in riverside marshes occurs up to the Great Glen and has been reported increasingly common in recent years along the lower reaches of the Spey. On stream banks the larvae of the butterbur moth *Hydraecia petasitis* feed on the roots of butterbur. This moth was for long regarded as restricted in Britain to a belt of country from northern England northwards to the Highland Boundary Fault, but it is now known to be much more widespread.

The larvae of some species of true flies are aquatic. Those which are found in Highland streams include various non-biting midges and some flies which do bite: the black flies *Simulium* spp, known in America as buffalo gnats because of the humped appearance of the thorax. The interesting point about this genus is the differing habitat requirements of its species. All of them need aerated water but some are found in slow-moving water; others favour the middle reaches where the current is strong, whilst others prefer the rapids near the source. Various species of mayfly larvae live in these fast-flowing streams, especially those belonging to the group of flat nymphs which are adapted for clinging to stones in strong currents. Free-swimming nymphs also occur, however, and in this category is *Baetis rhodani* which is widespread throughout the Highlands in the middle reaches of rivers. An arctic-alpine mayfly *Ameletus inopinatus* has been recorded from the source of the river Endrick. The caddis-flies present are mainly of those families whose larvae construct wide-mouthed silken tunnels to attract their prey, but small numbers of case-inhabiting larvae of the family Limnophilidae frequent the quieter lower reaches. Stone-flies are especially characteristic of swift-flowing mountain streams and in the upper reaches the nymphs constitute an important element in the invertebrate life during the spring.

Water-beetles adapt themselves to certain specialised habitats within the stream. On the rocky or gravelly bottoms of mountain streams the small beetle *Oreodytes rivalis* is fairly plentiful and another beetle of this genus *O. borealis*, an insect of more specifically northern distribution, is locally common. Several species of the family Helmidae, notably *Helmis maugei*, live under stones in the burns. These tiny beetles cling to the underside of stones and are nocturnal in habit, feeding on vegetable matter. Common free-swimming beetles include *Deronectes depressus elegans*, *Hydroporus palustris* and *Platambus maculatus* which live among the aquatic plants on the river margins. The surface-dwelling whirligig beetles *Gyrinus* spp are locally distributed.

Like many other aquatic insects, water-bugs occupy different levels and areas of a stream. Shore-bugs live on the edges of the water: where the margin is silt, *Salda littoralis* may be found; and where it is shingle, there may be *Saldula scotica*, a northern bug which, despite its specific name, is by no means restricted to Scotland. Most lesser water-boatmen inhabit still water but some are found in slow-flowing Scottish lowland rivers; two which often live in association are *Sigara dorsalis* and *S. falleni*. The water-scorpion *Nepa cinerea* prefers still water but it can tolerate the lower reaches of some Scottish rivers, for example the Tweed and the Endrick, provided that some vegetation is present. Its range, however, does not extend to the far north. The water-cricket *Velia caprai* lives on the surface and, unlike the water-scorpion, appears to require water devoid of plant life. Where pondskaters occur, they are usually restricted to back-waters.

There is only one dragonfly which is really typical of Highland rivers and that is the golden-ringed *Cordulegaster boltonii* which is fairly widespread, although there are gaps in its distribution. The nymphs will be found in deep pools where rock barriers have ponded the water. The nymph of the northern emerald *Somatochlora arctica* lives in boggy streams and a few damsel-flies haunt the quiet stretches of rivers.

The mollusc of most interest must surely be the pearl-mussel *Margaritana margaritifera*. The habitat of this bivalve is the gravel bed of a clear swift-flowing stream and it is found, sometimes in abundance, in a number of such streams in the west and north-west. Small pearls form in some of the shells and they are still fished by Scottish tinkers who use a glass box to enable them to see the bottom clearly. The pearl-mussel can live for as long as seventy years and an unusual characteristic is that it is probably the only mollusc in Britain which appears to prefer non-calcareous water. Two common and widespread molluscs in Highland streams are the river limpet *Ancylastrum fluviatile* which fastens itself to stones in the river-bed and the pea-shell cockle *Pisidium casertanum* which lives in sand or mud; but snails are not plentiful in this type of river and are found mainly in the lower reaches. The fountain bladder-snail *Physa fontinalis* requires clear running water and it lives in some Scottish streams where its other essential requirement, aquatic vegetation, is met. Slow-flowing rivers in southern Scotland have the flat ram's-horn *Planorbis complanatus*, the keeled trumpet-snail *P. carinatus* and *Bithynia tentaculata*. The Forth and Clyde Canal, as might be expected, is a rich hunting ground possessing many more species than do the Highland streams.

The greatest variety of fish live in the rivers of southern Scotland where some have been introduced. The distribution of fish in a particular river can to a certain extent be correlated with the three main zones—lower, middle and upper reaches—although some species are not necessarily restricted to one zone. The lower reaches are the haunt of coarse fish and of certain estuarine species. The two commonest fish here are the pike and the perch which are widespread although neither reaches the far north. The perch is beautifully coloured and those who wish to make its acquaintance may be fortunate enough to see one, as I once did, in the observation room at the fish-ladder at Loch Faskally in Pitlochry. The roach is another fish which is fairly common in the slower-moving larger rivers, especially those of southern Scotland. Other and scarcer relatives of the roach are the dace, which in Scotland is known only from the lower reaches of the Tweed; the common bream, small numbers of which are reported from the rivers Annan and Clyde; the chub, which is present in a few of the rivers of Galloway; and the gudgeon, which has been introduced into the Teviot and the Tweed. Those strange fish, the sea and river lampreys, live in the lower reaches and estuaries of a number of rivers in southern and central Scotland. The habitat of the small third member of this family, the brook lamprey, extends farther upstream and includes burns and even ditches. The sea lamprey visits rivers for the purpose of breeding, but the flounder breeds in the sea and spends the rest of the year in the estuaries and lower stretches of the rivers.

The middle reaches have a different assemblage of fish. Of the very small fish, the three-spined stickleback is well distributed throughout the country but the ten-spined is rare. The minnow likes clear rivers with a good depth of water and is believed to be native as far north as southern Banffshire, although introductions may have been made at some places farther north. It takes a careful scrutiny of the gravel bed to spot the stone loach, for its mottled scales match the gravel perfectly, as it rests motionless, waiting for its prey. This fish and the three-spined stickleback can tolerate a measure of pollution and can survive in stretches where trout have been ousted. The little bullhead is scarce but has been recorded from some streams in the Central Lowlands. Eels are generally distributed not only along the length of rivers but also in all types of still water; their greatest numbers, however, are in the middle reaches where they remain until the breeding urge takes them to the spawning grounds deep in the Atlantic.

The fish par excellence of this section of the rivers are those of the salmon family. Most of the large rivers are renowned for their salmon. Many of these fish, having arrived in the early spring, spend much of the summer in the middle reaches before moving upwards to the spawning-beds. The Hydro-Electricity Board has spared no effort to mitigate any adverse effects which its projects might have had on the salmon. Mention has already been made of the fish-pass at Pitlochry: it is 1,020ft long and consists of a series of gradually ascending pools connected by underwater pipes which enable the fish to negotiate the 60ft high dam. A different method is in use at Loch Awe, where an electrically operated lift every three hours raises salmon and sea-trout over the 30ft high dam. The disease which affects salmon was still of widespread occurrence in 1971 and research so far had failed to identify the causative organism. Both sea-trout and brown trout are abundant, moving to the upper reaches to spawn. In some of the Southern Upland rivers such as the Ayr, Nith, Clyde and Tweed and also in the Tay, with its tributary the Isla, the grayling has been introduced and does well.

The various habitats of the waterside attract an assortment of birds and many species find suitable nesting sites. The mute swan in Scotland is found mainly on lochs, both sea and freshwater, but some occur on the large slow-flowing rivers, and on the upper part of the Tweed early in June I have seen a family with five well-grown cygnets. It was on this river that a Bewick swan, which is very scarce even as a winter visitor, summered in 1958. The little grebe is another bird which in Scotland breeds mainly on still water, but some pairs nest along the quiet stretches of the larger rivers where moorhens and mallard also will be found. Sandy river-banks have colonies of sand-martins in such places as the Merse in south-east Scotland where the streams drain a basin of Old Red Sandstone, and on hill burns where there are banks of glacial sand. The other well-known bank-nesting bird, the kingfisher, is very scarce but small numbers breed in the Southern Uplands, mainly in Galloway, where there has been an increase in recent years; isolated birds are occasionally reported from the rivers of the Grampians. Where reed-beds adjoin the river, reed-buntings and sedge-warblers nest. Those elegant birds, the grey wagtails, can be seen along the whole length of a river. They are most frequent in the mid to upper reaches, although there are reports of their becoming scarcer in the Central Highlands during recent years. Pied wagtails are more numerous; though seldom far from water, they are not so closely dependent on it as are the grey wagtails. Shingle banks are the nesting sites of oyster-catchers, which are as much at home far inland on a streamside as on coastal shingle; ringed plovers and common sandpipers are other waders of this habitat. Statuesque herons fish the rivers, an occasional cormorant travels upstream and ospreys fish the Spey and the Tay. Goosanders frequent the upper reaches of many Highland streams, but they are much persecuted because their staple diet is fish: one wonders what the position will be when the ospreys become a little more numerous.

If there is one bird more than any other which the nature-lover associates with the fast-flowing mid and upper reaches of Scottish streams from the Borders to Sutherland, it is the dipper. As he stands on a mid-stream boulder jerking himself up and down, and even more as he indicates his mastery of his environment by walking under the water on the rocky bed, he is the river personified.

The scene changes in winter with the disappearance of the summer migrants and the arrival of considerable numbers of wildfowl on the lower sections of the major rivers. In particular, the Kirkcudbrightshire Dee, the Tay with its tributaries in Perthshire and the Deveron in Banffshire, have a large winter population. Mallard, teal and wigeon are the three common surface-feeding ducks and the diving ducks are the tufted and the golden-eye; the last-named in recent years has remained until late in the spring. The Tay has hordes of pink-footed and greylag geese; and in Galloway the Dee has, in addition, Greenland white-fronted geese. Whooper swans frequent many of the rivers, in contrast to the situation in the southern part of the British Isles where Bewick swans greatly outnumber whoopers.

Mammals of rivers are few and not a great deal is known about their distribution. The little insect-eating water-shrew is probably local over much of the mainland; it is reported as occurring commonly in the faunal region of the Clyde and it is present in the limestone streams of Sutherland, in various west-coast rivers and in Morayshire. Water-voles like habitats with dense bank vegetation and are probably most plentiful on the lower stretches of the larger rivers; but they have been noted from the small hill streams of Aberdeenshire and like the water-shrew they are probably of local distribution throughout the country. Melanism is quite common in the Highland race of the water-vole. Although the greatest density of otters is along the West Highland coasts, this mammal is widespread and fairly common on rivers and burns throughout Scotland. Otters travel far along burns and ditches from loch to loch, but due to their nocturnal habits they are seldom seen, except by the quiet angler, although spraint deposited on boulders provides evidence of their presence. In recent years the alien mink has spread rapidly across the country; it is now present in all mainland counties except Nairn, Sutherland and Caithness. Between 1962 and 1970 the campaign of the Department of Agriculture and Fisheries resulted in the killing of 2,000 mink. The situation is now considered to be under control, but it has to be accepted that complete eradication of this fierce carnivore is quite impossible and, like the grey squirrel in England, it is now a permanent member of the British fauna.

LOCHS

Much of Scotland is dotted with innumerable lochs, which vary in size from a few square yards to over 20 square miles. Many of these lochs have resulted from the actions of glaciers during the Ice Age which scooped out the corrie lochans high on the mountainsides, gouged out the deep rock basins at lower levels and eroded relatively shallow hollows, known as kettle-holes, in boulder clay or other fluvio-glacial soils. Some, such as the small dhu-lochans of the north-west, are formed from hollows in the peat; whilst a few on the Cambrian Limestone have been caused by the dissolution of the calcareous rock. Others are either artificial in origin, having been constructed for fishing or for ornamental purposes by the damming of streams, or have been greatly enlarged beyond their natural size by hydro-electric dams.

The longest freshwater loch is Loch Awe which is almost 25½ miles long, the deepest is Loch Morar with a maximum depth of 1,077ft and the largest is Loch Lomond with a surface area of nearly 27½ square miles. The larger lochs are not always of great depth and Loch Leven, for example, although over 5 square miles in area, seldom exceeds 20ft in depth. Many of these lochs make a notable contribution to the beauty of the Scottish scenery and the best viewpoint for the surrounding countryside is often from a boat in the centre of one of them.

In contrast to the rivers, the lochs contain a variety of plant communities with a certain degree of imprecise zonation. Abundant vegetation develops usually only on fine soils in sheltered bays, and many lochs have large stretches of open water devoid of plant life. Large stands of the common reed *Phragmites communis* occupy the fringes of many lochs, growing either as a pure community or in association with the small bur-reed *Sparganium minimum* or the shore weed *Littorella uniflora*. The reed-grass *Phalaris arundinacea* is not so well able to sustain the exposed conditions of Highland lochs but it is not uncommon in Lowland lochs; at Hoselaw Loch in the Tweed valley it grows with reedmace *Typha latifolia* and the very local umbellifer, cowbane *Cicuta virosa*. Those Lowland lochs which have nutrient-rich conditions sometimes have reed sweet-grass *Glyceria maxima*, yellow flag *Iris pseudocorus* and bogbean *Menyanthes trifoliata* growing in marshy areas at their edges. On gravelly soils in shallow water the common spike-rush *Eleocharis palustris* is often the dominant plant. This is accompanied sometimes by the bulbous rush *Juncus bulbosus* and shore-weed, the last-named extending into the zone of the water horse-tail *Equisetum fluviatile* which can commonly be seen in rather deeper water, sometimes with the bottle-sedge *Carex rostrata*. A characteristic and widespread plant of acid Highland lochs and the peaty lochans of the north-west is the water-lobelia *Lobelia dortmanna*, whose delicate lilac-coloured flowers open in July on erect stems arising from submerged leaves. Other plants occurring along loch margins include the true bulrush *Scirpus lacustris* and the fen sedge *Cladium mariscus*.

Scotland has several rare aquatics. *Juncus nodulosus* is a rare rush closely allied to the alpine rush *J. alpinus*. It had not been noted for some time at its original site at Loch Ussie in eastern Scotland, but it has been rediscovered there recently. It is known also from a site near Braemar. The flexible naiad *Najas flexilis* is a submerged aquatic which grows in Perthshire and has recently been found in Loch Kindar in Kirkcudbrightshire. Pipewort *Eriocaulon septangulare* has been known for some time from the Hebrides, but it was not until 1967 that it was discovered on the mainland at Ardnamurchan. Loch Lomond has the very rare Scottish water-dock *Rumex aquaticus* which grows 6ft tall. In August 1969 the small *Crassula aquatica* was first discovered in Scotland; it was growing on mud in western Inverness-shire in a typical aquatic marginal community of shore-weed, water-lobelia, lesser spearwort and spike-rush. There is no reason to suppose that it is not endemic, and other stations may yet be located.

Farther out in the water are the plants with floating leaves: conspicuous among these are the water-lilies of which all five native kinds are found. The white water-lily *Nymphaea alba* is much more plentiful than the yellow water-lily *Nuphar lutea*. In some lochs in northern Scotland a smaller-flowered subspecies of the white water-lily *N. a. occidentalis* occurs and mainly in this region too is found the least yellow water-lily *Nuphar pumila*. Beyond the lilies are the pondweeds: the broad-leaved *Potamogeton natans* is very common and the slender-leaved *P. filiformis* is a typical species of Scottish coastal lochs. The deepest water occupied by vegetation has plant communities of quill-wort *Isoëtes lacustris*, Canadian pondweed and willow-moss. The perennial quillwort closely resembles an annual, the awl-wort *Subularia aquatica*, which is also found in the same type of nutrient-poor loch. Many lochs have numerous small islands and these often have a well-developed vegetation of willow, birch, rowan and even occasionally mature pine.

The larvae of a number of moths feed on the marshy vegetation of loch margins, but Aberdeenshire represents the northern limit of most moths which occur in this habitat. The aquatic larvae of the tiny china mark moths *Nymphula* spp feed on pondweed, bur-reed and yellow water-lily in southern Scotland, but the numbers thin out farther north. The caterpillar of the crescent moth *Celaena leucostigma* lives on fen sedge and flag iris and another more local moth is the small clouded brindle *Apamea unanimis*. A few species are more particularly associated with Scotland. The Crinan ear moth *Amphipoea crinanensis* was first discovered in 1899 at the Crinan Canal and subsequently determined as a species in its own right, distinct from the ear moth *A. oculea*; the larvae feed on flag iris and have been found in various places in Scotland and northern England. The gold-spot *Plusia festucae*, whose caterpillars feed on flag iris and various sedges, is consistently plentiful in northern Britain up to the Great Glen.

Many kinds of water-beetle have been recorded from the lochs. Lowland lochs with dense aquatic vegetation have the largest numbers and the greatest variety of species. Morton Lochs in Fifeshire have the surface-living whirligig beetle of coastal waters, *Gyrinus caspius*, as well as the ubiquitous *G. natator*. The smallest member of this genus is appropriately named *G. minutus* and is more plentiful in the northern part of Britain, being especially typical of small moorland pools. An exclusively Scottish species is *G. opacus* which is found only north of the Central Lowlands. Dytiscid beetles are well known to pond 'dippers' and there is a Scottish species *Dytiscus lapponicus* of northern and western Scotland, found particularly in the smaller lochans. The genus *Agabus* has two species characteristic of Scottish mountain lochs, *A. arcticus* and *A. congener*; and other species of high altitude include *Deronectes griseostriatus* and *Acilius sulcatus*. In contrast to some other forms of insect life, there are quite a number of water-beetles which are more common in Scotland than in England.

A good selection of the common water-bugs live along the loch margins, particularly in the small, peaty moorland pools; and in addition there are some which are specifically northern in distribution. The pondskater *Gerris costae* frequents peat-pools at moderately high altitude; I have seen it in a tiny pool on the slopes of An Teallach in Wester Ross. *G. littoralis* is a pond-skater of northern Europe which is widely distributed throughout Scotland where loch margins have a silty soil. Upland pools are also the habitat of three mountain species of water-boatmen: *Sigara scotti*, *S. wollastoni* and *Arctocorisa carinata*. *S. wollastoni* has a pale form, which has been recorded from several lochs including Loch Leven and has been named *S.* var *caledonica*. A rare shore-bug *Saldula vestita*, first identified from specimens collected from the shores of Loch Leven in 1874 and then lost to sight for many years, was again found at the same loch in 1945.

Adult dragonflies range over a number of habitats, sometimes far from water, but since the nymphal state is always aquatic it is convenient to consider them in this section. Only three of the large hawker dragonflies can be seen in Scotland and of these only the common aeshna *Aeshna juncea* is plentiful and found in most counties. The blue aeshna *A. caerulea* is a northern species which encircles the world, but in Britain it is found only in some four Highland counties and it is usually far from common even there. The third hawker is the golden-ringed one already mentioned in the section on rivers. The medium-sized dragonflies known as the darters are a little better represented. The black-legged form of the common sympetrum *Sympetrum striolatum nigrifemur* flies in most of the counties of the western seaboard, and a from intermediate between this and the common species occurs in a few eastern counties. The only other sympetrum normally seen in Scotland is the northern species, the black sympetrum *S. danae*, which breeds in boggy pools and lochs with dense vegetation and is found commonly throughout most of the country. Two rare arctic-alpine dragonflies are the white-faced dragonfly *Leucorrhinia dubia*, which occurs in a few widely separated areas; and the northern emerald *Somatochlora arctica*, which also has a scattered distribution. The brilliant emerald *S. metallica* has a discontinuous distribution, occurring in southern England and in two Highland counties. The well-known four-spotted Libellula *Libellula quadrimaculata* is widespread and locally common, but the rest of the darters are seldom seen. Of the seventeen species of damsel-flies, only seven have been recorded from Scotland and some of these are rare. The large red *Pyrrhosoma nymphula*, the common ischnura *Ischnura elegans* and the common blue *Enallagma cyathigerum* are probably the three most plentiful damsel-flies. There is one arctic-alpine species, the northern coenagrion *Coenagrion hastulata*, which is very local in Argyll and Inverness-shire.

No reference to the invertebrate life of lochs would be complete unless brief mention is made of the molluscs which inhabit them. Southern Scotland has more kinds than the rest of the country and common species include the wandering snail *Limnaea pereger*, dwarf pond-snail *L. truncatula*, ear-shaped pond-snail *L. auricularia*, fountain bladder-snail *Physa fontinalis*, valve-snail *Valvata piscinalis*, the river-limpet *Ancylus fluviatilis* which despite its name is not confined to rivers, and several kinds of trumpet-snail *Planorbis* spp.

We are not concerned in this book with fish from the angler's viewpoint and therefore it will be sufficient to say that trout are abundant in the lochs and that in the larger waters they reach a great size. The native trout have been supplemented in many lochs with introductions of different kinds. Loch Leven has produced its own subspecies, a beautiful fish said to be of exceptionally good flavour and this has been introduced widely elsewhere. The small Loch Mhaolach-Coire at Inchnadamph in Sutherland has a population of trout which resembles a kind otherwise found only in Ireland and known as the gillaroo. Pike and perch are the most numerous coarse fish. There are roach in many lochs of southern Scotland and eels are widespread, but other fish are either scarce or localised. Bream, carp and tench are very local and rudd is known only in Culzean Castle Loch in Ayrshire.

The origin of the vendace is a mystery. The only place in the world where they are found with certainty is the small Mill Loch at Lochmaben in Dumfriesshire where they occur in some abundance; they were also at one time in the nearby Castle Loch but are believed to be extinct there now. They are elusive fish and are often not seen or caught over long periods. Studies were made during 1966 by P. S. Maitland of Glasgow University who, by using echo-sounding apparatus, found that the fish were concentrated in a few deep parts of the loch, rising only at dusk to feed through the night on surface plankton. The tradition that they were introduced by Mary, Queen of Scots is almost certainly incorrect; they are much more likely to be relicts of the Pleistocene era. A subspecies, the Cumberland vendace, is found in the Lake District and the powan of Loch Lomond and Loch Eck, already mentioned in Chapter One, belongs to the same genus. The char too has been mentioned in Chapter One, because of its Ice Age origin. Of the several races, that in Loch Rannoch is the most attractively coloured: it has a black back shot with metallic blue and a claret-coloured belly with pink-spotted flanks. Most of the char are located in Highland lochs and the only ones in the Southern Uplands are in three lochs in Galloway. In 1968 salmon disease was responsible for the death of a number of char in two lochs and it is thought that they may be particularly susceptible to this disease. The American brook-trout is misleadingly named for it is not a trout but a char. It has been introduced into a few Scottish lochs but I understand that it has not done well. In 1955 a new subspecies of char, with partly yellow colouration, was located in Loch Eck: it has been named *Salvelinus alpinus youngeri*.

In the bird life of Scotland, lochs play an important part. The larger ones are often situated in valleys which act as flight lines for migrating birds; and the lochs themselves serve as staging posts at which flocks of waterfowl will stay for a time before continuing to their winter quarters. Many of the lochs in winter receive wildfowl in large numbers. In summer too they often concentrate the bird life and their environs have a greater density than has the adjacent moorland. There is variety in the breeding species on individual hill lochs, as a walk across a lochan-studded moor will soon show.

Coots and moorhens are widespread at the lower levels but in general they are plentiful only south of the Highland Boundary Fault; the latter species is found more especially in small pools and reedy lochans. Mute swans have spread and increased during the present century, aided by various introductions during the last 150 years. Like the three preceding species, mallard and little grebe are common except in the north-west. Great-crested grebes now nest on a number of larger lowland lochs, the first breeding in Scotland being on the Loch of the Lowes in Perthshire in 1877. Grebes tend to be colonial nesting birds, and there are small colonies of two rarer species on a few lochs. The Slavonian grebe breeds on the margins of certain lochs in the Central Highlands and in northern Scotland, and in 1973 it bred for the first time on a Perthshire loch. The black-necked grebe is more restricted still, being found merely in one or two Central Lowland lochs, and only very occasionally will a pair or so breed in lochs elsewhere. In addition to mallard, teal, wigeon, shoveller, pochard and tufted duck nest in a number of counties; but common scoter, gadwall and pintail are much more localised, and scaup nest but seldom. In 1971 the nesting of a pair of golden-eye in Inverness-shire was confirmed, and breeding had probably taken place the previous year. It looks as though this species, which has for some years been staying longer into the spring, may become a regular breeder.

Several of the larger water-birds in Britain nest only on Scottish lochs. Notable among these are the divers. The red-throated diver is the smallest and builds its nest on the small lochans,

Page 125 (*above*) Typical scenery of the north-west, Torridonian Sandstone resting on Lewisian Gneiss: Lochan an Ais in the foreground, Cul Beag in the left middle distance and Stac Polly in the background; (*below*) red stag on a Perthshire hillside

Page 126 (*above*) Female fox moth resting on sand-dunes at Invernaver, Sutherland; (*below*) the mountain avens, a plant of limestone here growing near sea-level on the north Sutherland coast

mainly in the north-west. These lochans are not able to provide the birds with food and although the birds could fish the larger lochs, they prefer to flight to the open sea where an abundant supply of fish is readily available. The typical, hunchbacked silhouette, fast flight and guttural croaking are a familiar sight and sound in Wester Ross and Sutherland. The red throat is not a conspicuous feature since it is a dull maroon, and in cloudy conditions, from a distance, it appears black. The black-throated diver is a rather larger, and I consider a more handsome, bird with a superb pattern of black and white stripes on the side of its neck. It nests usually on the large lochs, often those with islands, on the edge of which the birds can nest undisturbed. The breeding territory is similar in range to that of its smaller relation, but it does not nest so far south, Rannoch Moor probably being the most southerly breeding area.

The divers' habit of nesting almost at the water's edge is of great help to them in effecting a speedy departure on the water, since they have considerable difficulty in walking on land; but it is hazardous in that a slight rise in the water-level will often destroy the nest. The birds certainly try to combat such a disaster by a swift second attempt at nesting: a ghillie has told me of eggs smashed in a storm and of the hen's being found incubating two more eggs ten days later; and it has been suggested that, in a similar instance, the bird must have moved its eggs to another place when the rising water-level flooded the original site.

Some non-breeding great northern divers remain late in northern Scotland. The first proved breeding in Britain took place in 1970 on the mainland, although in the past breeding has been suspected in the Shetlands. No nesting was known in 1971 and it is early yet to determine whether or not the breeding was an isolated occurrence or the start of a regular trend. Canada geese were introduced into Scotland in the eighteenth century and now breed in Wigtownshire and the Central Lowlands. Feral greylag geese in Galloway have spread and increased from stock introduced at Lochinch in Wigtownshire about 1930; and these birds are now breeding in the wild state throughout the three south-western counties. Smaller numbers breed in northern Scotland, and although their status is somewhat obscure, these have derived at least partly from genuine native stock. A few whooper swans will stay through the summer, especially if injured; and occasionally a pair will breed. This may happen rather more frequently than is generally supposed owing to the remoteness and inaccessibility of many of the northern lochs.

One of the most exciting ornithological events in Scotland this century was the return of the osprey. This bird has had continued successful nesting despite many hazards. In the eighteenth and early nineteenth centuries it was fairly common, nesting as far south as Galloway, with a preference for ruined lakeside buildings such as Inchmahome Priory on the Lake of Menteith and Loch an Eilean castle, where the last nesting was in 1899. There is no doubt at all that egg-collectors, procurers of birds for taxidermy and gamekeepers harried this unfortunate species to extinction. A pair bred on Loch Arkaig in 1908 and there is what seems to be a reliable record that a pair nested on an island on Loch Loyne in 1916. From then on the osprey was extinct as a breeding species. However, in the early 1950s, ospreys began to be seen during the summer and in 1955 an eyrie was located late in the season at the Sluggan Pass in Inverness-shire. On 13 March 1959 an area around Loch Garten, where the birds had now settled, was designated a statutory bird sanctuary with entry forbidden during the breeding season. From 1958 onwards the Royal Society for the Protection of Birds has organised a continuous watch each summer; but even so, eggs have been stolen more than once and the enterprise has had to endure the additional hazards of storm damage, forest fire and unsuccessful breeding. All in all, however, the project has been a success and great credit is due to those whose bold enlightened policy has combined protection of a rare species with opportunity for the public to observe a magnificent bird and to receive a lesson in nature conservation. By 1971 seven pairs were breeding, and at the Loch of the Lowes the Scottish Wildlife Trust has also constructed a hide for the use of the public. Osprey-watching is proving to be a major tourist industry, and in 1972 no less than 100,000 people visited the Loch Garten eyrie.

Large colonies of common and black-headed gulls nest on lochsides and on islets. Their breeding sites are not restricted to the lochs but can be found also on moorland bogs. The former species is plentiful and well spread throughout the country, except in the south-east. The latter species is generally abundant south of the Highland Boundary Fault; farther north it is fairly common in the Central and Eastern Highlands, but distinctly less plentiful in the west. Of the wading birds the common sandpiper is a familiar sight round the shores of many a loch, piping plaintively and bobbing anxiously up and down. The Temminck's stint is a rare wading bird anywhere in Britain, even on migration, so that a nesting pair is an outstanding event. A pair nested on Speyside in 1934 and 1936 and possibly also in 1935 on the shore of a large loch. A single bird was seen during the summer of 1947 and breeding was attempted in 1956. A pair was seen in courtship display in 1969 and breeding was proved in 1971 in a northern county.

Large flocks of waterfowl assemble on the lochs and reservoirs and spend the winter there. Of course, not all the aggregations will have the same specific composition, and there is a considerable amount of movement as lochs become frozen over and the ducks descend to lower levels and to the coast. Many of the large waters in the south and east have numbers of greylag and pink-feet. Bean geese have become very rare, but a few birds are usually seen each winter in Galloway and at one or two places along the east coast.

The various amphibians—frogs, common toads and newts—are widespread and the natterjack toad has colonies along the Solway coast and in Renfrewshire. No description of Scottish lochs would be complete without reference to the monsters that are alleged to haunt some of them, especially Loch Morar and Loch Ness. Of these two the latter is, to my mind, much the more important. Legends and traditions are associated with certain lochs and no doubt many can be rightly regarded as fairy tales of no concern to a naturalist. A considerable body of evidence, however, has been amassed that something unusual lives in Loch Ness. This evidence is the more impressive in that some of it has come from eyewitnesses who were previously sceptical. I believe that the right approach, indeed the scientific attitude, is to keep an open mind until more substantial evidence has been obtained.

CHAPTER SEVEN

Woodland

Mixed deciduous woodland–Birch woods–Beech, ash and alder woods–Coniferous forests

MIXED DECIDUOUS WOODLAND

THE NATURAL VEGETATION climax of the Southern Uplands, the Central Lowlands and the glens and lower slopes of the Western Highlands is deciduous woodland in which oak is the predominant species. How this immense area of woodland was destroyed has been outlined in Chapter One. Today only a few relict semi-natural woods remain, although these have been augmented in southern Scotland by a certain amount of planting both of shelter belts and of more extensive areas in the policies belonging to far-seeing landowners. Sessile oak *Quercus sessiliflora* is the native species of the wetter, shallower soils of the west but the pedunculate oak *Q. robur* has been extensively planted on the deeper brown earth soils of the south and east.

To enter a sessile oak wood in a western glen is to observe many points of difference from a walk in an English pedunculate oak wood. To begin with, the ground consists mainly of rocky outcrops instead of deep clay soil. These rocks and the ground between are densely covered with mosses, one of the commonest and most conspicuous of which is *Rhytidialelphus loreus* with its long red stems and large curved leaves; there is a rich bryophytic flora including liverworts. The heavy rainfall of this region produces a humid atmosphere, which is conducive also to the epiphytic growth of mosses and ferns such as the common polypody *Polypodium vulgare* on tree-trunks.

The herb layer includes grasses such as the tufted and wavy

hair-grass *Deschampsia caespitosa* and *D. flexuosa*, sweet vernal grass *Anthoxanthum odoratum* and soft grass *Holcus mollis*, whilst ferns are represented by a number of species, including bracken *Pteris aquilinum* which covers large areas. Grazing often greatly reduces the numbers of the common woodland plants, and dog's-mercury *Mercurialis perennis* is not found in the abundance so evident in English woods. The presence of such diverse plants as erect tormentil *Potentilla erecta*, common cow-wheat *Melampyrum pratense*, woodruff *Galium odoratum* and sanicle *Sanicula europaea* is indicative of the varying nature of the underlying rocks. Damp woods have lesser celandine *Ranunculus ficaria*, ramsons *Allium ursinum* and yellow pimpernel *Lysimachia nemorum*. In this habitat also is a very local flower, the yellow star-of-Bethlehem *Gagea lutea*, growing in scattered localities between Ayrshire and Inverness-shire. The very rare whorled Solomon's seal *Polygonatum verticellatum* grows in a few woods in the Southern Uplands, and in Perthshire, principally within the area of the Tay river system.

The shrub layer is often sparse, due in the main to the intensity of the grazing. Rowan *Sorbus aucuparia* and holly *Ilex aquifolium* are fairly common on the siliceous soils of the west. The typical cherry of these northern woods is the bird-cherry *Prunus padus*, but the gean *P. avium* also occurs. Although oak is dominant, these woods have a mixed composition, often with much birch: *Betula pendula* in the drier parts and *B. pubescens* on the wetter soils. This climatic division means that the former predominates in the east and the latter in the west. Other associated trees are the aspen *Populus tremula*, wych- or Scots elm *Ulmus glabra* and, in the damper parts, alder *Alnus glutinosa* and willow *Salix* spp.

Not many butterflies find the northern climate congenial. Green-veined whites *Pieris napi*, ringlets *Aphantopus hyperantus* and small pearl-bordered fritillaries *Boloria selene* are probably the most widely distributed of the woodland butterflies. The pearl-bordered fritillary *B. euphrosyne* is fairly widespread but is not known in the far north. The purple hairstreak *Quercusia quercus* can be seen only in the oak woods of southern Scotland

and the speckled wood *Pararge aegeria*, which is such a common English insect, is very local, occurring in scattered localities of the west and rarely north of the Great Glen.

There is a greater variety of moths and a few are northern species. The moth known as the plain clay *Eugnorisma depuncta* could hardly have a more uninspiring name, but its appearance is rather more attractive than its name implies. The larvae feed on a variety of plants and the moth flies in the woodlands of the eastern Grampians, where it is probably more plentiful than anywhere else in Britain. The Saxon moth *Hyppa rectilinea* has been recorded from a few places in the Southern Uplands, but it is definitely more common north of the Highland Boundary Fault. The chrysalis of the lunar marbled brown *Drymonia ruficornis* and the merveille du jour *Dichonia aprilina* are buried among the roots at the base of oak tree-trunks. Many other species occur, most of them more plentifully in the southern part of the country. Wych-elm is a fairly common constituent of the tree layer and two moths in particular are associated with it. The larvae of the dusky-lemon sallow *Xanthia gilvago* and of the brick *Agrochola circellaris* feed on the wych-elm seeds; the former is recorded as fairly common in Berwickshire while the latter is more widespread wherever its chosen tree grows.

Many spiders live in woodland and some are equally at home in deciduous and coniferous types. Some of the commonest are *Oonops pulcher* and *Drassodes lapidosus* under stones, *Lepthphantes minutus* and *Drapetisca socialis* in tree-trunk cavities and *Meta segmentata* in the foliage. The cave-spider *Meta menardii* has been reported from a hollow tree in Hamilton High Parks in Lanarkshire, and Dr E. A. Crowson has suggested that this may have been the species which attracted the attention of Robert Bruce in that famous encounter in a cave.

The sparsity of the shrub layer in the sessile oak woods of the north-west means that summer warblers like the blackcap and garden-warbler, which nest in dense cover, are absent. They are local breeders in the pedunculate oak woods of the Southern Uplands and the Central Lowlands, but they are scarce north of the Highland Boundary Fault. The garden-warbler is one of several species which are slowly spreading northwards, and the chiff-chaff is another which has now reached the northern counties. It has been suggested that, for the chiff-chaff, rhododendron is an important habitat factor. Two birds which the English birdwatcher will surely miss are the nightingale and the nuthatch which do not normally reach Scotland at all, although there have been occasional sightings of the latter in various places. The birds which can be expected with certainty in almost any area of oak or mixed broad-leaved woodland are chaffinch, robin, wren, song-thrush, blackbird, starling and various tits, in particular the blue tit. In the far north where nesting sites for tits are scarce, I have seen a blue tit feeding young in the top of the iron post of a passing-place notice. Willow-warblers are likely to be present, although they will not be as numerically important a component as in the birch woods. The wood-warbler's requirement of mature deciduous woodland with negligible shrub cover is met in many parts of Scotland. This is another bird which has spread northwards during the present century and has reached the northern coast. It breeds now in every county and this increase contrasts with an apparent reduction in numbers in many English counties. That handsome bird, the redstart, nests in mature oak woods or wherever there is decaying timber. The first recorded nesting of the marsh tit in Scotland was in the summer of 1945 in Berwickshire, although these birds had been known there for a considerable time before this and had probably nested. Marsh tits are still largely restricted to this county, but they have been seen in recent years in Roxburghshire and Selkirkshire. The main Scottish breeding area of the pied flycatcher is in Galloway and Dumfriesshire, but scattered pairs nest northwards to Inverness-shire.

The lesser-spotted woodpecker does not occur and the great-spotted woodpecker, which became extinct or almost extinct in the middle of the last century, began recolonising only just before the beginning of the present century. There is a little evidence that the species may not have entirely died out in certain remote areas of eastern Scotland but at any rate the recolonisation came not from this nucleus, if such existed, but northwards across the Border. This steady but unexplained spread brought nesting birds to Midlothian by 1901 and to Argyllshire by 1919; breeding was proved in Inverness-shire in 1920 and Sutherland was reached by 1926. During this time and since, the birds spread out in all directions to occupy suitable habitats in uncolonised areas between the focal points. When Baxter and Rintoul published their mammoth work on Scottish birds in 1953, they recorded the green woodpecker as an uncommon visitor; but it now breeds sparingly in the Southern Uplands and the Central Lowlands and the species is being seen well north of this region up to Wester Ross. Although there are no nuthatches, tree-creepers are fairly common. An astonishing event in 1969 was the breeding of three pairs of wrynecks in Inverness-shire after birds were seen occasionally between 1951 and 1968. Even in southern Britain the bird is on the western edge of its range and its great decline and contraction of breeding area in these southern counties are well known. It is all the more remarkable, therefore, that breeding should take place in Scotland so far to the north, and it may have originated from a westward displacement of a few birds which had been migrating northwards to Denmark or to southern Norway. It is early yet to determine whether or not the wryneck can be added to the list of regular breeders.

Of the Corvids, jackdaws are common, as are carrion and hooded crows in their respective zones. Jays have an irregular distribution up to the southern Grampians, with some areas of suitable habitat unoccupied; and it is reported that some ravens have returned to tree-nesting in south-west Scotland. Rooks and magpies will be referred to in the chapter on farmland (see page 164), as will wood-pigeons and stock-doves. The tawny owl is widespread over most of the country, but the little owl is a recent arrival as a breeding species; it now nests regularly in an area of south-east Scotland and single birds are occasionally seen elsewhere. The northward spread has been slow and uncertain, but there seems no reason why it should not eventually establish itself. Birds of prey such as the buzzard, sparrow-hawk and kestrel all suffered a decline in recent years, the first-named primarily due to myxomatosis in the rabbit and the last two coincident with the increased use of toxic chemicals; but a slow recovery has taken place and in 1971 all three species did well. Of the larger ground-nesting birds, pheasants are increasing in mixed woodland; woodcock are largely restricted to damp woods where they nest in relatively small numbers over most of the country, being scarcest in the north. I have, however, heard them on the Stoer peninsula roding over the tops of the trees.

Few if any mammals are restricted to deciduous woods, but one or two are more typical of this habitat than of others. The badger undoubtedly prefers broad-leaved woodland and it is local but increasing in these situations from the Border to Sutherland. The fieldmouse is plentiful and especially common in dry oak woods. A few mammals, on the other hand, are more characteristic of other habitats but are nevertheless found in these woods, either as permanent inhabitants or temporarily whilst searching for food. Such are the graceful roe deer which browse in these woods and the wild cat, an animal of moorland mountain slope which lives also in the woods, particularly in scrub on steep rocky slopes. Bank-voles are more woodland creatures than are their relatives, the short-tailed voles; but both species occur. Other small mammals are moles and the two terrestrial species of shrew. Among the mammalian predators, fox, stoat and weasel traverse the woodland floor in search of their prey. Within the shelter of the trees they are themselves safe from other than human predation, but outside they sometimes fall victims to the eagle. Due either to deliberate introductions or to accidental escapes, grey squirrels now live in the wild in various parts of Scotland. A pair was released in Dunbartonshire in 1892 and within fifty years their progeny were occupying a large block of country north-eastwards into Perthshire. Edinburgh has a population of grey squirrels originating from

Corstorphine Zoo. Just across the waters of the Forth, Dunfermline in Fifeshire was the site of another introduction in 1919. Individual squirrels have been seen at several places in the Southern Uplands and as far north as Loch Shiel. The grey squirrel's preference for broad-leaved woods may help to restrict its spread in a country where only a relatively small proportion of the total woodland area is deciduous. Red deer were originally woodland animals, so that it is not surprising that during the first part of the year they live in the shelter of the wooded glens. The fallow deer is not native to Scotland and those that live in the wild are the result of escapes from parks. They find a congenial habitat in parkland and mature deciduous woods and occur in scattered localities in several counties.

BIRCH WOODS

Birch is often co-dominant with oak in Scottish broad-leaved woodland and a common constituent of pine woods, so that to consider this species as a distinct habitat may be regarded as creating a somewhat artificial division. There are however two reasons which I feel justify such a separation: namely, that a considerable number of extensive birch stands exists both in the Grampians and in the Northern Highlands; and because birches have their own specialised fauna. There are two species in these woods, *Betula pendula* and *B. pubescens*, and their geographical distribution has been mentioned earlier in this chapter (see page 132). In actuality, the situation is rather more complex, for a great deal of hybridising takes place between the two kinds, and Brown and Tuley in a recent study of birch woods in Glen Gairn, Aberdeenshire found that no tree could be assigned to one species or the other. Yet the typical trees of each species are quite distinct in appearance: *Betula pendula* when mature forms a whitebarked tree with graceful pendulous branches and is aptly named the queen of the woods; *B. pubescens* has a more shrubby, upright outline and hairy twigs.

There is much similarity between the moss floras of birch and

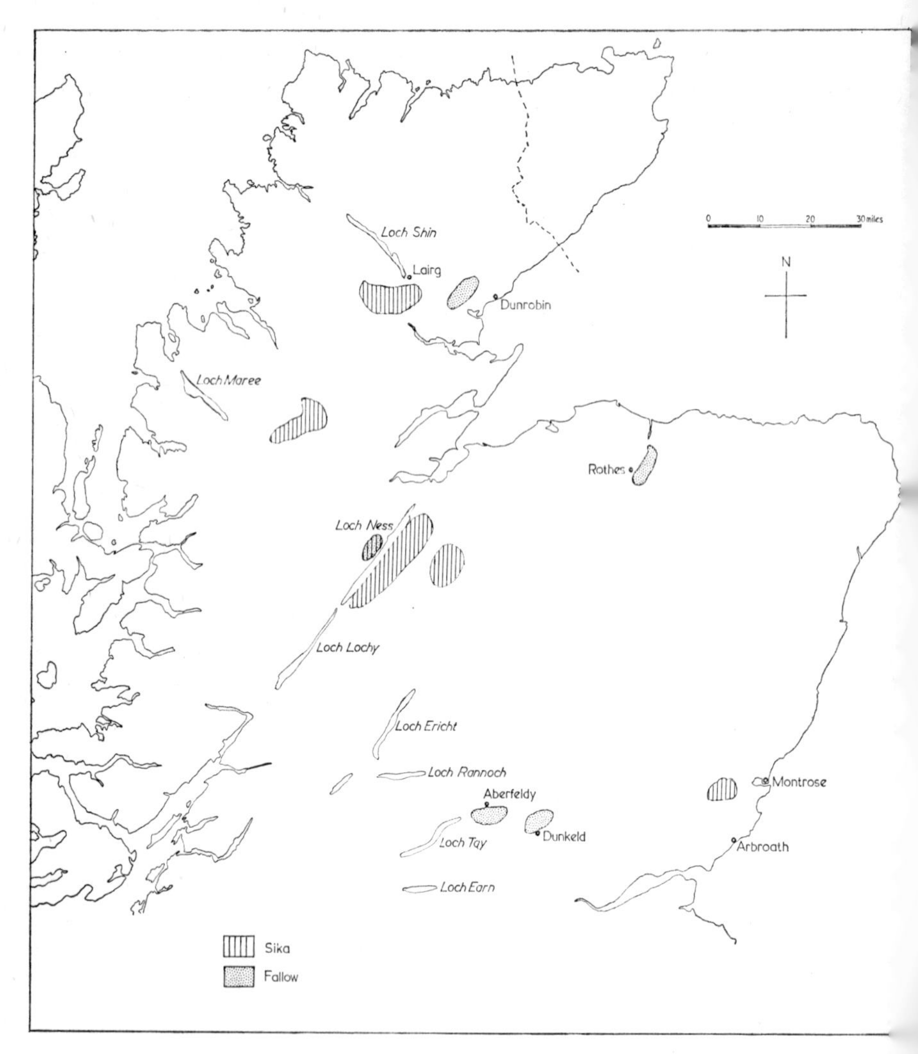

Fig 3 Map showing distribution of deer in the Grampians and North-west Highlands. Widely distributed throughout the region are roe deer in woodland, and red deer mainly on hill ground.

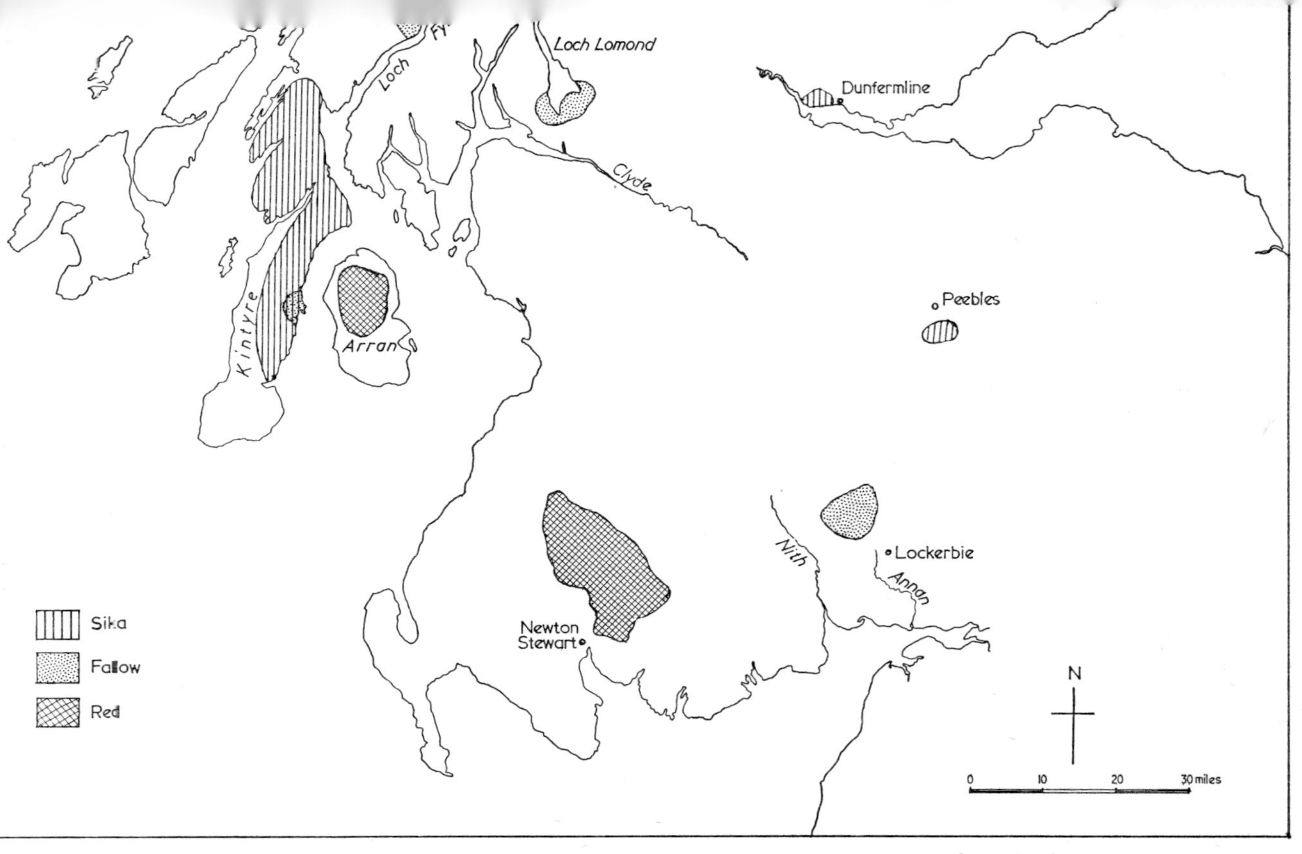

Fig 4 Map showing distribution of deer in the Southern Uplands and Central Lowlands. Roe deer are widely distributed in woodland on the mainland.

oak woods. On dry siliceous rocks birch woods have a very restricted calcifuge flora, but this sparsity is redeemed by the pleasing aspect of leafy branches shivering in the wind above a thick, rolling carpet of ling, bilberry and cowberry. Even degenerate birch wood has a certain bizarre appeal in the misshapen trees and fallen trunks clothed with *Polyphorus betulinus* and *Polystictus versicolor*. Heath bedstraw *Galium saxatile*, tormentil *Potentilla erecta* and wood-sorrel *Oxalis acetosella* are other flowering plants which grow in these heathy birch woods. On deeper, less acid soils there may be a considerable amount of grass cover, especially sweet vernal grass, and a richer variety of plants including primrose, common violet and mountain buckler-fern.

In addition to the few common butterflies which may be seen in open areas of birch there is one species which is of particular interest. This is the chequered skipper *Carterocephalus palaemon*, an extremely local butterfly, which was found only in a few woods of the English Midlands until 1942, when numbers were seen flying in an area near Fort William. The habitat is open birch wood and, as might be expected from the food preference of the larvae, the herb layer is dominated by grasses. The caterpillars feed on brome-grass from July to October, then hibernate, and the imago flies in June sunshine. It is apparent that this colony in Inverness-shire is a race of its own, distinct from that in England. It is possible that the former race somehow survived the final glaciation, but the remarkable fact is that this insect should have survived so long undetected in what is a major tourist region.

A considerable number of moths are associated with birch. The lesser swallow prominent *Pheosia gnoma* is a spring-flying moth, more common in northern Britain. The yellow horn *Achlya flavicornis*, the pebble hook-tip *Drepana falcataria* and the scalloped hook-tip *D. lacertinaria* also fly in the springtime; they are common and widespread in the Central Highlands and except for the last-named occur in the Southern Uplands as well. Summer-flying moths include the large emerald *Geometra papilionaria* and the silvery arches *Polia hepatica*. The larva of the former species has the ability to change its colour to match the surroundings; it hibernates in a red-brown colour resembling that of the twig on which it rests, but in the spring its colour changes to the green of the birch leaves on which it feeds. The silvery arches is a local moth in Scotland which is perhaps most frequent in the eastern half of the country. The orange underwing *Archiearis parthenias* and the argent and sable *Rheumaptera hastata* can be seen flying around birch trees in the sunshine; the latter is especially characteristic of the northern birch woods in Sutherland. Two specialities belong to this habitat. The Rannoch sprawler *Brachionycha nubeculosa* is a rare moth of elegant appearance which occurs on birch in the neighbourhood of Loch Rannoch and at Aviemore. The Kentish glory *Endromis versicolora* is a larger and more beautiful moth, still of open birch woods at various places as well as in the two localities mentioned above, although it has not been recorded from Rannoch in recent years. The sexes differ considerably from each other: the male is smaller, has hind wings of a rich tawny brown and flies in the April sunshine, whereas the female is larger, has hind wings with a white background and is a nocturnal flyer.

A small blue leaf-beetle *Cryptocephalus parvulus* is a rather local species which feeds on the leaves of young birches in some areas. A member of the family of click-beetles, *Corymbites impressus*, is a northern species which lives on scattered trees of birch and pine at their higher altitudinal limits. F. Fraser Darling and J. Morton Boyd record the occurrence of the bee-beetle *Trichius fasciatus* in the birch woods of the Eastern Highlands. This is a rare insect mainly centred in North Wales and is so named because of its hairy body and black-banded orange elytra which give it a superficial resemblance to a bee. A number of plant-bugs live on birch including the birch shield-bug *Elasmostethus interstinctus* and a large flat-bug *Aradus betulae*, a species of northern distribution living under birch bark; in Britain this flat-bug is known only from the southern Grampians.

There is a fairly distinctive bird community in the Scottish birch woods, the core of which consists of willow-warblers, chaffinches, tree-pipits, wrens, robins and coal-tits. Other tits rather less common are great, blue and long-tailed. The opportunist bullfinches, lesser redpolls, thrushes, blackbirds and spotted flycatchers are other fairly common birds. The great spotted woodpecker in its northward colonisation has occupied a number of birch woods as far north as Sutherland. An interesting ornithological development in recent years has been the establishment of the redwing as a breeding species in northern Scotland where it nests often, but not exclusively, in birch scrub. The first known nest was discovered in 1925 in Sutherland and in the next sixteen years birds bred on several occasions in various northern counties. After a gap of twelve years when no nests were discovered, more breeding pairs were found; and it appears that the numbers are increasing with about fifty pairs nesting in 1971 from Inverness-shire northwards. An even rarer bird of this habitat is the brambling, which has nested in northern Scotland on a few occasions during the present century.

BEECH, ASH AND ALDER WOODS

On a much smaller scale are a few other types of broad-leaved woods. Small areas of beech have been planted on the more fertile soils along the roadsides; on Forestry Commission ground; and as shelter belts on farmland, especially on the large estates in the Southern Uplands, Central Lowlands, Argyllshire and the Eastern Highlands, particularly in Aberdeenshire. They succeed surprisingly well, reproducing themselves freely even in the exposed north-east; and mature trees, whether by the shore of Loch Sunart in Argyllshire or flanking an Aberdeenshire road, contribute much to the beauty of the landscape. The tallest beech-tree known in Britain, some 142ft high, grows on an East Lothian estate. At Dawyck near Peebles in the Southern Uplands a remarkable fastigiate form of the common beech was discovered during the last century. The 600yd long beech-

Page 143 (*above*) Wild cat at entrance to cairn; (*below*) young Highland fox in Perthshire

Page 144 Golden eagle. Note the immense talons

hedge at Meikleour in Perthshire is well known; planted in 1746, the fateful year of Culloden, it is now over 90ft tall. Beeches find their most congenial situation on the valley floor or in other sheltered places, but they will grow at levels around 1,300ft. The herb layer often consists of soft grass; anemone *Anemone nemerosa*; wood-sorrel *Oxalis acetosella*; wood-stitchwort *Stellaria nemorum* in wet conditions; wood-speedwell *Veronica montana*, a very local flower on boulder clay; great wood-rush *Luzula sylvatica* on acid soils; and wood-sanicle on basic soils.

Drapetisca socialis is a spider often seen on the trunks of beeches and the harvestmen *Oligolophus tridens* and *O. hanseni* are recorded as fairly common on beech bark in Aberdeenshire. The northward spread of the wood-warbler brought it in 1949 to the beech wood at Tongue in north Sutherland. This particular wood also has a colony of badgers, descendants of a pair which escaped from captivity in 1925. A winter visitor to beech woods is the brambling, the numbers of which fluctuate considerably from year to year; it is only in good beech-mast years that these birds visit Scotland in large numbers.

Ash woods are few and woods which consist predominantly of this species are found only on the ancient Cambrian Limestone in the north-west. The best-known one is the 30-acre Raasal wood, a National Nature Reserve in Wester Ross. I have visited this wood at the time when it presents its most attractive aspect, with wild hyacinths and primroses in flower. As well as the vernal flowers the ground layer includes bracken, slender false-brome *Brachypodium sylvaticum*, melancholy thistle *Cirsium heterophyllum*, meadowsweet *Filipendula ulmaria* and yellow pimpernel *Lysimachia nemorosom*. The shrub layer consists of scattered hazel, rowan and hawthorn. The dominant bird of these Scottish ash woods is the chaffinch, with blackbird, robin, wren, willow-warbler, wood-pigeon and various tits commonly found.

Alder woods are more numerous, on moist hill slopes, on burnsides and on the alluvial floor of the glens. Apart from such stands as these which have alder as the dominant tree, alder-birch woods and mixed woods with alder as a component have developed in some parts of the country. Flowering plants include marsh-thistle *Cirsium palustre*, marsh-hawksbeard *Crepis paludosa*, golden saxifrage *Chrysoplenium oppositifolium*, marsh-marigold *Caltha palustris* and many other moisture-loving species. Rowan is an associated tree in the drier woods and the hairy birch and sallow *Salix atrocinerea* in the wetter. The iron prominent moth *Notodonta dromedarius* was reported as widespread in alder woods in Berwickshire although it has apparently declined there in recent years: the specific name derives from the four humps on the back of the caterpillar. The moths in Scottish localities are darker in colour than those obtained in England. The common birds of woodland are found in summer, whilst flocks of siskins and redpolls are typical of the winter scene. The willow-tit nests in this habitat principally in south-west Scotland, although it is not restricted to alder and willow; it is probably most numerous in the valley of the Clyde. K. Williamson states that some of the redwings of north-west Scotland are nesting in coppiced alders.

CONIFEROUS FORESTS

Seven per cent of Scotland's surface area, 1,500,000 acres, is covered with conifers; and the area increases each year. Most of this consists of commercial plantations of the Forestry Commission and of private landowners. One-third of Britain's afforested land is in Scotland, and in 1971 the Forestry Commission had 875,000 acres of plantations in 170 Scottish forests. The three most commonly planted trees are Scots pine *Pinus sylvestris*, spruce *Picea* spp and larch *Larix* spp. Although the Scots pine is the only native of the trio, it is not by any means suited to all areas of Scotland. The Forestry Commission has found, for example, that it does not do well on the Silurian rocks which cover such a wide area of the Southern Uplands. Norway spruce *Picea abies*, the traditional Christmas tree, has been grown at least from 1682 onwards, when it was known to be growing at the head of Loch Fyne. This too cannot be planted anywhere; it does best in fairly sheltered conditions on good soils in the valleys. Another member of this genus is the Sitka spruce *P. sitchensis* which, even at a distance, is easily distinguished from its better known relative by the blue tinge to the needles and, on handling, by the sharper points to the needles. This tree is more accommodating in its requirements and is therefore extremely useful to the forester for planting on the higher, more exposed levels and on the less fertile soils. It is extensively grown in the Southern Uplands, where it is often used as shelter belts; but in very exposed situations in the north-east, the Sitka has itself to be protected in the early stages of growth by an alpine pine *Pinus mugo*. This, the common mountain pine on the continent, is able to tolerate extreme conditions of climate.

Numerous other conifers, including Douglas fir *Pseudotsuga menziesii*, are grown and the visitor will notice the prevalence of larches, particularly on the vast wooded estates around Dunkeld, Pitlochry and Blair Atholl. It was from these woods in 1897 that there originated the hybrid between the Japanese and European larches known as the Dunkeld larch *Larix eurolepis*, which is a more vigorous grower than either of its parents and able to do well even on infertile soil. Wind damage is an ever-present hazard in the plantations during winter and spring, not merely to trees past their prime but more seriously to trees in the pole stage. In recent years 1,000 acres have been affected in Loch Ard Forest alone and some 250,000 trees have been blown down in 650 acres of Knapdale Forest.

A small proportion of the total woodland consists of scattered remnants of the native Caledonian pine forest. The old woods in the Spey and Dee valleys, at Rannoch, on the northern slopes of Ben Eighe and in Glen Affric are among the largest; but relict pine woods exist also in Perthshire, in the Great Glen and in various other areas, mainly in the west and north. The principal difference between a plantation and a natural wood is that whilst the former is composed of even-aged trees, the latter comprises trees of various age groups, although areas where a complete succession occurs are rare. The old native woods also are generally of a much more open character, but there is no essential difference between the plant and animal communities of such a wood and those of a mature plantation, so the two types will be dealt with collectively.

In contrast to the paucity of flowering plants, mosses in variety occupy the ground layer and are often luxuriant over the rocks. This is especially so in the wetter climate of the west; and the plantations of the Argyll and Queen Elizabeth Forest Parks, for example, have mosses in greater abundance than have the native pine woods and plantations of the Eastern Highlands. Even so, some dry parts of the old pine forests in the Spey and Dee valleys have a field layer in which a few mosses, including *Hypnum cupressiforme* and *Hylocomium splendens*, are dominant often to the exclusion of other forms of plant life.

Apart from on the verges flanking the rides and forest clearings where a number of plants grow, the flowers of a plantation in the thicket and pole stages are very few; indeed, where the cover is dense, the carpet of needles combined with the shade excludes all flowering plants. In a mature wood of more open character there may be much wavy hair-grass, ling, bilberry and cowberry. Much of the native pine wood is of a more open nature still, and here may be found a few rare or local northern species belonging to this habitat.

The commonest of these is chickweed-wintergreen *Trientalis europaea*, a small plant of the Primula family which is not very conspicuous, since it grows singly, but its white flower on a short stem springing from a whorl of leaves is a common sight during June in such places as the Black Wood of Rannoch and the Speyside woods. The intermediate wintergreen *Pyrola media* is a very local plant, occurring mainly in Scottish pine woods; but it grows also on moorland and I have found its white globular bells among bell-heather on the moor south of the Black Wood of Rannoch. The common wintergreen *P. minor* is also local, but in moist parts of the Culbin Forest in

Morayshire it is quite common, as is its relation, the serrated wintergreen *Orthilia secunda*. Probably the rarest member of this family is the one-flowered wintergreen *Moneses uniflora* which is almost entirely confined to some pine woods in eastern Scotland from Perthshire northwards. It is generally rare but is reported to have increased rapidly in Culbin Forest. In Scotland it bears the charming local name of St Olaf's candlestick. *Linnaea borealis*, a plant of northern pine woods, is quite rare and found only in the east of the country. Its altitudinal limit is about 2,000ft, where it grows in birch scrub; and its stations on moorland probably indicate the site of earlier woodland. A trailing plant, its pendulous pink flowers are very beautiful and it is reputed to have been the favourite flower of the great Swedish botanist, Carl Linnaeus, after whom it has been named.

Two uncommon orchids of northern distribution are found in some Scottish coniferous forests. One is the creeping lady's-tresses *Goodyera repens*. The other is the coral-root *Corallorhiza trifida* with yellowish-white flowers and a fantastically shaped rhizome which, however, is certainly not to be dug up for inspection. It is one of the saprophytic species, lacking roots and leaves and at least partly dependent on soil fungi for its existence. This orchid is locally plentiful in some moist coniferous woods, especially in the east. A third orchid of this habitat which is relatively common is the lesser twayblade *Listera cordata*. It requires an open situation within a wood and occurs also on moorland. But these botanical treasures are bonuses, for anyone visiting a native pine wood such as that of Rothiemurchus for the first time could surely find satisfaction enough in the springy ground beneath his feet, in the resinous scent inhaled with every breath and in the sunlight slanting through the parasol pine-tops to create an ever-changing pattern on the shaggy red-brown boles.

Large shrubs are absent from the plantations but the native woods have small tree species such as rowan, bird-cherry and especially juniper *Juniperus communis*. At Tynron in Dumfriesshire the Nature Conservancy has on a steep slope an unusual wood which consists almost entirely of juniper. This shrub grows in a variety of forms: some hug the ground; others spread outwards with straggling branches to make rather untidy bushes; and others develop a fastigiate habit. In deer-forest country juniper is a valuable source of food for red deer in wintertime.

There are no butterflies which are especially associated with coniferous forest and few species inhabit the forests; but green-veined whites are seen flying in the plantations and small pearl-bordered fritillaries are fairly common in the old open woods like those at Glen Feshie and Rothiemurchus. Conflicting statements have been made about the status of the pine hawk-moth *Hyloicus pinastri* in Scotland but it is apparently not known to exist in Scotland at the present day. The pine beauty *Panolis flammea* is reported as widespread among pines in south-east Scotland, and it is probably generally distributed throughout the country except in the most northerly part. Bordered white *Bupalus piniaria* can be seen in numbers, flying in daylight around pine-trees during June, both in the native woods and in the plantations. Since World War II this moth has reached pest proportions and the larvae have caused considerable damage by defoliation in some east-coast forests. The Rannoch looper *Semiothisa brunneata* is a distinctively Scottish moth, first discovered in the early part of last century in the entomologically famous Black Wood of Rannoch. The caterpillars feed not on pine but on bilberry leaves under the trees. The pine carpet *Thera firmata*, the grey pine carpet *T. obeliscata* and the spruce carpet *T. variata* are common in the plantations. Two pug moths which feed on juniper are the juniper pug *Eupithecia pusillata* and the Edinburgh pug *E. intricata* whose vernacular name belies its widespread distribution.

The prevalence of huge conical heaps of pine-needles in the coniferous forests indicates the abundance of the wood-ant *Formica rufa*. Wood-ants have no directly adverse effect on forestry, but the same cannot be said about some other members of the order Hymenoptera. Imported materials often introduce to the country undesirable insect aliens: one such has been a saw-fly *Anoplonyx destructor*, which feeds on larch and has become very plentiful in the large areas of larch in Central Scotland. In another insect order, aphids can be troublesome in plantations and Sitka crops in Scotland have suffered heavy attacks from these pests.

Beetles of many kinds abound in the woods, far too many for the foresters' peace of mind. Some of course are harmless, like the ground-beetles scurrying over the woodland floor or the burying-beetles such as the one I watched in Craigvinean Forest, Perthshire: this was attempting to bury a large black slug whilst encumbered itself with more than twenty orange mites. Two longicorn beetles *Rhagium bifasciatum* and *R. mordax* are found in decaying timber and in tree-stumps but not in healthy pines. They are recorded as widespread in Scotland, and I have seen the former even in scattered pines high up Glen Einich in the Cairngorms. Some other Longicorns live in growing trees but do not generally cause serious damage, certainly not as much as a number of weevils and Scolytids which sometimes assume plague proportions: *Hylobius abietus*, *Hylastes ater* and *Pissodes pini* are common pests. An examination of the effects of the great wind-blow in 1953 on insect-pest populations was made by Crooke and Kirkland, who noted an increased breeding of the Scolytid beetle *Blastophagus piniperda* in fallen trunks, resulting subsequently in severe damage to trees which remained standing.

In recent years the small insects known as ambrosia beetles have caused considerable damage in coniferous forests in western Scotland. One of the genera in the family Scolytidae is *Ips*. The males in this genus are bigamous and after the mating cavity has been dug out and utilised the two females proceed to excavate their egg-laying galleries: one goes vertically upwards and the other straight down so that never the twain shall meet. Domestic harmony is ensured! A new species for Scotland has been added to this genus with the finding in north-east Scotland of *Ips cembrae* known as the larch bark-beetle. This has established itself and is reported to be spreading southwards, where it will certainly find a large area of suitable habitat in Perthshire.

Quite a number of the beetles in coniferous woods are northern species which, in the British Isles, are either confined to Scotland or at least found more plentifully there. They include the small Clavicorn beetle *Pityophagus ferrugineus*, locally common under bark; a rare and beautiful cardinal beetle *Schizotus pecticornis* resplendent with scarlet colouring and pectinate antennae which can be seen sometimes on old pine-stumps; and in contrast, a beetle which is found on young conifers, the tiny red weevil *Brachonyx pineti*.

The bird community of coniferous woodland varies in composition according to the stage of growth which the trees have reached. For a few years after a moor has been planted, moorland birds such as skylark, meadow-pipit, whinchat, stonechat and wheatear continue to breed. If, as sometimes happens, field-voles multiply to plague proportions, predators (in particular, short-eared owls) will move in and establish territories until the decline in the vole population causes them again to disperse. Such a vole plague occurred in 1952–3, in the Carron valley north-east of the Kilsyth Hills in the Central Lowlands, where thirty to forty pairs of short-eared owls nested on 3,500 acres; and this event was the subject of a detailed study by J. D. Lockie. Another predator which has established itself on Forestry Commission land is the hen-harrier. This fine bird of prey became extinct on the Scottish mainland towards the turn of the century, but it began to recolonise suitable terrain by the end of World War II and has spread to a considerable number of localities. Unfortunately, despite the Bird Protection Acts, hen-harriers are still subject to much persecution, especially on grouse moors. They have a discontinuous distribution from Galloway to Sutherland, perhaps attaining their greatest density in a belt across the southern Grampians.

As the trees continue to grow, the small woodland birds enter the young plantations. In the tit family, coal-tits are pre-eminently the tit of coniferous woodland, whether planted or native. In a few areas, mainly in the old Caledonian forests of Speyside and to a lesser extent on Deeside, but also in mature plantations such as the Culbin Forest, may be found the gem of this family: the crested tit. These birds are not abundant even in their chosen localities and have to be carefully searched for. The low, twittering call-notes, though difficult to describe, are highly distinctive and once heard will readily identify the birds on a subsequent occasion. Blue, great and long-tailed tits also occur but are much less common than coal-tits. Goldcrests are numerous and always seem to be *the* bird of Douglas firs, for one can be fairly certain of hearing the thin piping of goldcrests from any stand of these trees. Two members of the finch family which are typical of plantations are the siskin and the lesser redpoll. The former usually nests high in conifers and is fairly common in this situation in eastern Scotland, less so in the west. It is a most attractive little bird as it indulges in tit-like acrobatics, calling continually with its soft nasal 'scree' or fluttering in song flight high in the clearings. Lesser redpolls are not so restricted to a coniferous habitat for nesting purposes; nevertheless, they are noted for their habit of moving into young plantations. They are widespread but probably most common in western Scotland. The crossbill, like the crested tit, is a speciality of the old pine forests of Speyside, but it is not confined to native woods and the first one I saw was in a small plantation at Newtonmore. The species is plentiful in Morayshire and occurs from Wester Ross and south-east Sutherland south to Perthshire. Chaffinches, robins, wrens and bullfinches are other resident small birds; and hedge-sparrows nest in the shrub layers of juniper in the native woods. Song-thrushes and missel-thrushes are more common than blackbirds and the trunk-climbing birds mentioned in the section on deciduous woodland (see page 135) are found also in the coniferous forests. Summer migrants include tree-pipit, redstart and willow-warbler, which are all fairly common in the native woods.

The game-birds of coniferous woods are woodcock, pheasant, black game and capercaillie. Woodcock numbers are swollen by winter visitors. This is a bird whose breeding status in south-west Scotland has improved commensurate with the increase in plantations. It appears from W. B. Alexander's inquiry into the status of the woodcock in the British Isles in 1934 that the term 'roding' for the spring flight was then apparently not well known in Scotland. This is somewhat surprising, for although the etymology of this word is uncertain, it derives probably from either French or Scandinavian sources, countries with which historically Scotland has had strong links. Pheasants are particularly plentiful in the Southern Uplands, where also in Kirroughtree Forest near Creetown are numbers of golden pheasant.

Black game are birds whose ideal habitat is essentially the woodland edge where there is some shrub cover such as rhododendrons. It is therefore on the fringe of plantations that this species is likely to be encountered, but the open nature of the native woods in the Spey and Dee valleys greatly enlarges the extent of this habitat and black game are fairly numerous here. They feed principally on a variety of vegetable material and in winter they cause a certain amount of damage when they turn to the buds of conifers; but they also consume a quantity of insects, especially the heather-beetle *Lochmaea suturalis*. The gathering of blackcock for early-morning displays at 'leks' is well known and there are many of these sites in Scotland, often maintained over a considerable number of years. At one locality in Speyside an old bird occasionally still returns and displays at a long-abandoned site.

Striking in appearance though the blackcock is, it is overshadowed in magnificence by the cock capercaillie, the 'Cock of the Woods', which is half as big again and is the largest gamebird in Britain. Its distribution is confined mainly to the mature conifers of the Grampians, but it occurs in the coastal forests of Moray, and north to Easter Ross and the south-east corner of Sutherland. Capercaillie became extinct in Scotland by the middle of the eighteenth century. An attempted reintroduction in 1827 proved a failure, but ten years later a second attempt in a different locality at the eastern end of Loch Tay was more successful and by 1860 the capercaillie was firmly re-established in Perthshire as a breeding species once more. The cock is polygamous, and the spring calls are a succession of the most extraordinary sounds to come from a bird. Old birds are notoriously aggressive and have been known to attack people and vehicles.

The tawny owl is principally a bird of deciduous woodland, but small numbers breed in fir woods. It is reported to be the commonest owl on Speyside. The long-eared owl is the owl of the tall, dense coniferous woods, but it is a largely silent and elusive bird and its true status in Scotland appears obscure. It occurs in small numbers in the old pine forests and is probably locally distributed throughout the country. Kestrels are common, although conifers are not a favourite habitat for them. Sparrow-hawks, on the other hand, have a preference for large tracts of coniferous woodland. During World War II the sparrow-hawk increased in numbers, but after 1945 a disastrous drop in population ensued, reaching a nadir in the early 1960s. Greater control over chemicals and, to a lesser extent, the placing of the sparrow-hawk in the list of protected species, has resulted in a recovery which in 1971 was still continuing. The density varies in different areas and a recent woodland survey revealed 6 pairs per square mile in Dumfriesshire and only 0·5 pair per square mile in the Spey forests. The sparrow-hawk's larger relative, the goshawk, used to nest in Scotland up to about 1880 when it became extinct. In recent years there have been reports of goshawks in southern Scotland and it is likely that one pair at least are regularly breeding. The common buzzard nests in pine woods, among other places, but in general it must be regarded more as a bird of the hillside and mountain. The same applies to the golden eagle, pairs of which on occasion build their eyries in Scots pine.

The small mammals of woodland such as the common shrews, fieldmice, bank- and field-voles are widely distributed but not abundant. Bats are few in species, but long-eared, natterers and daubenton's bats are to be found more particularly south of the Great Glen. They are not of course restricted to conifers but occupy deciduous woods as well. Red squirrels have their strongholds in some of the mature woods, whether planted or native. They are scarce in many places and common only in the central and eastern Grampians. The British red squirrel is differentiated from those on the continent by its light-coloured tail. It is likely that most of the red squirrels at present in Scotland derive from various introductions made from the eighteenth century onwards, when the heavy disafforestation virtually extinguished the native population. The pine-marten is a predator of squirrels, but owing to the scarcity of these in the main marten areas, a variety of other prey is taken. Martens occur in the forests of the north-west, but they also live in open country and will be referred to in more detail in a later chapter.

In the Forest of Corriedoo near the small town of New Galloway in Kirkcudbrightshire, a cairn is reputed to indicate the spot where the last wild boar in Scotland was killed. It is of interest, therefore, to note that only a few miles south-west of this forest, a wild boar and sow escaped from a private estate in 1970, established themselves in the wild and bred eight piglets. The adults and all but one of the young have been shot. This one is in a plantation and can claim the proud title of the only wild boar in Scotland!

Roe deer are widespread in all types of coniferous woodland. They are mainly nocturnal creatures, but I have seen them feeding on the edge of woodland at midday in both the Spey valley and South Strome forest in Wester Ross. In the evenings they will emerge from the shelter of the woods to feed in the open fields. It was while watching roe in a hayfield near Loch an Eilean that I heard a strange piping that I could not identify coming from nearby trees: tracking the note to its source, I saw a roe fawn for the first time. Red deer live in the open areas of the old pine forests during the winter months when they are driven from the high tops; and they do damage by eating seedling pines, thus preventing regeneration. A few sika deer live in one or two of the Caledonian forest remnants in Wester Ross, but most occur in planted woodland. The greatest numbers are in Kintyre and on the shores of Loch Ness.

CHAPTER EIGHT

Farmland

Plants of field and wasteland–Invertebrates–Birds–Mammals

THIS SHORT OMNIBUS chapter deals briefly with those man-made habitats where direct human influence is constantly being exercised. The topography of the country and the nature of the soil ensure that farming is predominantly concerned with the rearing of livestock in its various forms and with the production of fodder crops. Hill sheep-farming occupies vast areas of land, no less than 7 million acres which is 35 per cent of the total land area. In the Southern Uplands it is the principal land use of the grass moors, but farther north it shares in multiple land use of grouse moor and deer forest with some of these latter habitats providing only summer grazing for the sheep. Dairy farming is practised in Galloway, Ayrshire, Lanarkshire, parts of the Central Lowlands and a few areas elsewhere but the rearing of beef cattle is more important, the major districts being the east-coast lowlands of Easter Ross and Inverness-shire, the Buchan plateau in Aberdeenshire, the Spey and Dee valleys, the Vale of Strathmore and the eastern part of the Southern Uplands. It will be noticed that the two categories of farming are divided on a climatic basis, the wetter western areas being more suitable for dairy cattle. Arable land of all types occupies only a relatively small proportion of the total land surface and much of this is cultivated for cattle food. The climate is too cool for much wheat production, but some wheat is grown on the Merse in Berwickshire and in the east-coast lowlands. Scotland is famous for its potatoes and these are grown in Ayrshire and the Lothians. Clydesdale and the district around Edinburgh are the main areas for horticulture and the Carse of Gowrie and parts of Angus are noted for their soft fruit, particularly raspberries.

PLANTS OF FIELD AND WASTELAND

The large rounded hills of the Southern Uplands are mainly grass-covered and carry heavy concentrations of sheep. To the motorists who pass through in their thousands, these hills appear to have a uniform vegetation; but this is not really the case, for there is often a marked altitudinal difference in the soils, with a corresponding difference in the composition of the grasses. The base may have a loamy soil enriched with minerals leached from higher up the slope and here sheep's-fescue and bent are the dominants; but farther up the hill the soil is thinner and there is often much peat supporting a grass-heath community with wavy hair-grass, mat-grass and purple moor-grass occasionally intermixed with ling. There is in addition a geographical variation with wet grassland of purple moor or mat-grasses more prevalent in much of the western part, whilst bent and fescue predominate in the east. The heavy grazing by sheep as well as by some wild mammals ensures that plants other than grasses are few in number of species. Low-lying grassland may have the beautiful globe-flower *Trollius europaeus*, various orchids such as the marsh-orchid *Dactylorchis incarnata* and the lesser butterfly *Platanthera bifolia*, heath-rush *Juncus squarrosus*, slender St John's wort *Hypericum pulchrum* and two rare plants in south-east Scotland: the yellow marsh saxifrage *Saxifraga hirculus* and the bird's-eye primrose *Primula farinosa*. A local flower of hill pastures is the umbellifer known as baldmoney *Meum athamanticum*, and others include thyme *Thymus drucei*, heath milkwort *Polygala serpyllifolia*, field-gentian *Gentianella campestris* and frog orchid *Coelglossum viride*. Dry hill pastures sometimes have sheep's-sorrel *Rumex acetosella* in abundance, with some of the flora of heathland including the moss *Polytrichum juniperum* and the tormentil *Potentilla erecta* which the shepherds of the Border hills know as the ewe daisy. Bracken encroaches on the grassland in places and often suppresses the grass, with a consequent loss of grazing; it is difficult to eradicate and is reported to be on the increase in the Lammermuir hills. Prolonged snow cover is conducive to the growth of a micro-fungus *Fusarium nivale*, which sometimes attacks bent-fescue grassland and causes considerable damage.

Arable fields of eastern Scotland have their quota of weeds, with the alien rayless mayweed *Matricaria matricarioides* in its preferred habitat of well-trodden ground. Despite increasing use of herbicides, corn-marigold *Chrysanthemum segetum*, corn-flower *Centaurea cyanus*, redshank *Polygonum persicaria* and poppy *Papaver dubium* still survive in the cornfields.

The roadside verges have a considerable variety of flowers and in some places there are great splashes of colour from a concentration of one species. The composition of this verge flora varies somewhat from region to region, and a selection of the plants which I noted when travelling in Scotland in 1973 will indicate the general pattern. In the Lothians the scarlet poppies towered above the carpet of white clover *Trifolium repens* and farther inland, deep in the Moorfoot and Lammermuir hills were the lacy umbels of hedge-parsley *Torilis japonica*. When Perthshire is reached, even the most disinterested of urban dwellers cannot fail to notice the glorious golden-yellow of the broom on the roadside banks. The verges in these Perthshire glens often have a dense vegetation which includes the wood cranesbill *Geranium sylvaticum* a local plant, red campion *Silene dioica*, greater stitchwort *Stellaria holostea*, crosswort *Galium cruciata*, lady's-bedstraw *G. verum* and, in shady waysides, wood avens *Geum urbanum*. The main road on the western side of Loch Tay is a splendid locality, with varying rock formation bearing a flora which includes meadow-vetchling *Lathyrus pratensis*, tufted vetch *Vicia cracca*, bird's-foot trefoil *Lotus corniculatus*, harebell *Campanula rotundifolia*, meadow-cranesbill *Geranium pratense* and many others. The showy ox-eye daisies *Chrysanthemum leucanthemum* come into the picture in Inverness-shire, around Fort William. Broom continues to flaunt its vivid colour through the counties of the north-east and poppies can be seen as far north as the Cromarty Firth. Here in this county of Easter Ross the great wood-rush *Luzula sylvatica* has strong societies by the roadside near Dornoch Firth; in the western part of Ross-shire rhododendrons *Rhododendron ponticum* are naturalised in some places. Foxgloves *Digitalis purpurea*, flags *Iris pseudocorus* and many ferns and mosses bedeck the coastal roads of western Inverness-shire and Argyllshire. In this latter county the regions of Knapdale and Cowal are well endowed with roadside flowers: ragged robin *Lychnis flos-cuculi*, red campion, beaked hawksbeard *Crepis taraxacifolia*, red clover *Trifolium pratense*, ox-eye daisies and the bamboo-like stems of Japanese polygonum *Polygonum cuspidatum*. The Tweed Valley branch of the Scottish Wildlife Trust has carried out a survey of the more uncommon plants of waysides in Berwickshire and indicated their siting to the Highway Authority, so that care can be taken when verge-trimming and other maintenance work is in progress. Hoary ragwort *Senecio erucifolius*, maiden pink *Dianthus deltoides*, blue sow-thistle *Cicerbita macrophylla* and *Aremonia agrimonoides* which is closely related to the common agrimony, are some of them.

Raspberries have naturalised themselves in many places in the eastern part of the country and now grow along railway embankments where, too, gorse and broom are frequent. The Oxford ragwort *Senecio squalidus* is rare in Scotland, but it has spread into the Southern Uplands and grows in profusion on a railway bank on the outskirts of Edinburgh. In the Central Lowlands the pit and ash bings which are a prominent part of the industrial scene provide a home for a miscellaneous assortment of native and alien plants, especially annuals.

INVERTEBRATES

Relatively few species of grasshoppers can tolerate the colder climate of Scotland and the largest populations are found in the grasslands of the Southern Uplands. One of the few which is well distributed over most of the country, except for a puzzling absence from the Buchan plateau, is the meadow-grasshopper *Chorthippus parallelus* which is at home in the lower-level grasslands. On the higher slopes it is replaced by the common green grasshopper *Omocestus viridulus* which occurs over almost the entire country up to north Sutherland. These are insects of moist grassy areas. In drier localities, such as roadside verges and wasteland, the dominant species is the common field-grasshopper *C. brunneus*; it is widespread in the Southern Uplands, but is more localised farther north, and is not recorded from the Northern Highlands at all. Certain introduced species are found in the man-made habitats. Of these the greenhouse camel-cricket *Tachycinis asynamorus* is indigenous in the eastern parts of Asia but has become established in Britain within the last hundred years. Its original introduction was probably by way of a hothouse plant and its habitat is restricted to heated greenhouses. The distribution of this insect in Scotland is therefore closely linked with the horticultural industry in Ayrshire, Lanarkshire and the Lothians.

Butterflies too are comparatively scarce; they are at their most plentiful in southern Scotland. That common butterfly of fields and wayside, the meadow brown *Maniola jurtina*, exists in a brighter-coloured form in Scotland: the subspecies *splendida*. Similarly the brown argus *Aricia agestis* had a Scottish subspecies which has now been promoted to full specific status *A. artaxerxes*. The wall *Lasiommata megera* is fairly common, flying over dry sunny wasteland in the more southerly counties; while in shady meadows are local colonies of the ringlet *Aphantopus hyperantus*. The orange-tip butterfly *Anthocaris cardamines* and the marsh-fritillary *Euphydryas aurinia* haunt marshy fields; neither reaches the northernmost counties and the latter is only locally distributed. Other butterflies of rough grassland are represented by the small heath *Coenonympha pamphilus*, common blue *Polyommatus icarus*, small copper *Lycaena phlaeas*, large skipper *Ochlodes venata* and, in a few localities in the Central Highlands, the dingy skipper *Erynnis tages*. The peacock *Inachis io* had been extinct in Scotland from about the beginning of the present century; but by the end of World War II it had become fairly common, if somewhat local, in the Southern Uplands and by the 1950s was well established in parts of the Grampians. Another butterfly of gardens which, similarly, may spread northwards is the holly blue *Celastrina argiolus*, which had not been known in Scotland until after World War II when it was observed in Dumfriesshire.

Grasslands have their quota of moths. Mat-grass is the preferred food of the caterpillars of the feathered Gothic *Tholera decimalis* in the fields of the Southern Uplands and the eastern lowlands. The caterpillars of the antler moth *Cerapteryx graminis* sometimes reach plague proportions on grassland. Rough grassland below 1,000ft is the haunt of the autumn-flying, slender-striped, rufous *Coenocalpe lapidata*, a distinctly northern species in Britain. The town of Paisley gave its name to a melanic form of the grey pug *Eupithecia subfuscata*, known as the Paisley pug, which is found occasionally in the valley of the Clyde. The masses of broom in eastern Scotland are the food of moth larvae of several species; and broom is also the food plant of the larvae of a small relict population of the frosted yellow *Isturgia limbaria*, which lives in Wester Ross.

The yellow hill-ant *Acanthomyops flavus* is common in the southern half of the country, but distinctly rare in the north; red ants *Myrmica* spp and the black lawn-ant *A. niger* are fairly common on farmland. On dry grassland where ragwort grows, the small beetle *Longitarsus jacobaeae* may be found; it has been reported abundant in the meadows skirting Loch Broom in Wester Ross. The common beetles of cultivated ground, such as the ground-beetles of the family Carabidae and the click-beetles of the family Elateridae, are widespread; in the latter family several members of the genus *Corymbites* are of northern distribution. Purple moor-grass is the preferred food of *Stenodema holsatum*, a capsid bug abundant on hill pastures; *S. calcaratum* is a related species which is more plentiful in the drier grasslands.

The upland pastures carry an enormous population of spiders, harvestmen and mites. A. B. Roy carried out a survey (published in 1955) of the spiders at the 1,000ft contour on a part of the Pentlands and estimated that there were about 800,000 per acre. The vast majority of these are tiny spiders belonging to the large family Linyphiidae. The generally uncommon *Erigone longipalpis* of this family has been recorded as plentiful on grassland in Aberdeenshire. The so-called harvest-bug *Trombicula autumnalis*, actually a mite, is reported to have become much more numerous in Fifeshire during recent years. It is the larval stage of the harvest-mite which can cause irritation to human beings.

BIRDS

The common birds of farmland are found in Scotland as elsewhere, but there are some regional differences. The house sparrow, which is so taken for granted south of the Border, and which is now numerous in Scotland, has not always been so. It was only locally common until the beginning of this century, when a considerable increase took place. At the present time it is most abundant in the south, especially in the urban areas, but it is rather scarce in mountainous districts and on some upland farms. It is difficult to realise that the ubiquitous starling was almost extinct on the mainland of Scotland in the first thirty years of the last century. Then began one of those northward spreads which several species have experienced, and by 1850 scattered colonies had reached Aberdeenshire. From this time onwards expansion accelerated and by the beginning of the present century the starling was well established almost everywhere.

Two other typical and abundant birds of farmland are the rook and the wood-pigeon. Rooks are common and well distributed throughout the country, wherever there are suitable habitats. Probably the greatest density is attained in the four counties of the north-east lowlands and there in the mixed woodland surrounding Hatton Castle near Turriff in Aberdeenshire is the largest rookery in the British Isles and perhaps in the world; the numbers of course fluctuate from year to year, but a census in 1957 revealed 6,697 nests in 2,570 trees. The neighbouring county of Banffshire claims the second largest rookery in Britain at Tore of Troup, with over 3,000 nests; and in south-west Scotland, Dumfriesshire has several rookeries with over 1,000 nests each. The wood-pigeon, too, prefers a habitat combining wood and farmland. The continued increase in the number of Forestry Commission plantations has favoured the wood-pigeon for, much more than the rook, it nests in coniferous trees, especially in dense stands of spruce, where communal breeders like the rook could not so easily establish nesting sites.

The fortunes of the stock-dove have fluctuated widely. It was not known in Scotland until about 1866; but from that time onwards it spread northwards, mainly along the eastern part of the country, up to Easter Ross, but also to some extent in the south-west up to Argyllshire. A decline began in the late 1950s, affecting even the north-east lowlands where it had been very numerous; but some recovery has now taken place and in 1973 it appeared to me to be fairly common in the north-east. It is strange to find the magpie so scarce, when it is all too much in evidence over most of England. Up to the early nineteenth century it was plentiful enough in suitable habitats in most of Scotland; but it is now mainly found in the south-west, although of local occurrence in the east from the Lothians to Easter Ross.

Several waders breed on farmland. The lapwing is a very common and widespread bird on grassland, particularly on marginal farmland, and it is increasing in some localities. The curlew is another well-distributed species, especially on the grasslands of the Southern Uplands and Aberdeenshire; Baxter and Rintoul state that the ploughing campaign of World War II did not oust curlews, which then nested among oats. Snipe nest most often on boggy moor or marshland, but they are by no means infrequent on rough pasture. Oyster-catchers are generally associated with shingle, whether on the shore or river side, but in some areas they regularly nest in arable fields. Gulls spend much time on farmland where, following the plough, they are a familiar sight.

Two distinctive birds of arable land are the cornbunting and the corncrake. The preferred habitat of the former is in coastal cornfields; this bird has declined in recent years but is locally distributed throughout much of the country, being most numerous in eastern counties, especially in Aberdeenshire. The corncrake has suffered a much greater decline in the present century and on the mainland is now found principally in the north-west, but there are welcome signs of a small recovery in the Central Lowlands and the north-eastern counties. Corncrakes are not restricted to cornfields; indeed, they are not especially characteristic of this habitat. Along the western seaboard they are typical of marshland and rough pastures and I have heard them calling from potato patches on crofting land.

Pheasant and common partridge are fairly well distributed over most of Scotland, though scarce in the north; and quail have nested sporadically in Scotland during 'quail' years. The barn-owl declined greatly in the last century. It has made a recovery in some areas, and I have found it commonly in Galloway, but it is reported scarce in south-east Scotland. Skylarks and meadow-pipits are common on cultivated land; the various finches of this habitat are rather more localised, being plentiful in some districts and scarce or even absent from others, but in winter large flocks congregate on lowland farms. At this season flocks of duck and geese graze in the fields, the different species having individual food preferences. For example, in Perthshire, it has been observed that wigeon feed on plant stems, mallard on various seeds and teal on small invertebrates in the wetter fields. The gorse of waysides provides nesting sites for yellow-hammers, linnets and stonechats, and the large fields of raspberries in the Central Lowlands attract numbers of seed-eating birds in August. Urban habitats throughout the country have been colonised by the collared dove, the spread of which did not follow any consistent pattern. The first arrivals in Scotland were seen in 1957 at Girvan in Ayrshire, and in the same year birds, which were possibly immigrants from Norway, arrived in Morayshire. Numbers continue to build up.

MAMMALS

There is nothing distinctive about the mammals of Scottish lowlands unless one includes the herd of wild white Cadzow cattle at Hamilton High Parks in Lanarkshire. Hedgerows and gardens provide shelter for small mammals and mole-heaps are a common sight in the fields. The brown hare is numerous on arable and grassland over most of the country, but is less common in the north-west. Although typically a mammal of the lowlands, it has a range whose upper limit is about 2,000ft, so that at the higher levels it overlaps the territory of the mountain hare. Rabbits have survived successive attacks of myxomatosis, and since the initial onslaught of this disease it has been recorded from Lanarkshire that they have changed their habits and moved into suburban gardens. They provide one of the principal sources of food for the stoat.

Stoats and weasels are widespread on farmland. I have never seen so many weasels anywhere as in the Southern Uplands, where they can be observed even on the roadsides. There are conflicting reports on the relative abundance of weasels and stoats. The fact that weasels seem to be more frequently sighted than stoats does not necessarily mean that they are more plentiful. It may be because weasels inhabit hedgerows and farmyards, whereas stoats range over a wider and higher terrain. When one of the periodic plagues of short-tailed voles occurred in Scotland in 1892, the Government instituted a Committee of Inquiry which came to the conclusion that the main predator was the weasel and that every effort should be made to conserve its population. Foxes are well distributed on cultivated ground, although in the north-east lowlands they are reported to be less plentiful on low ground than in the hill country. Much suffering has been caused to wild animals by the use of the gin trap, which was permitted in Scotland after it was outlawed in England; but now the Agriculture (Spring Trap) (Scotland) Act has prohibited its use from 1 April 1973, and it is hoped that this will have some restraining effect.

CHAPTER NINE

Moor and mountain

Heather moors—Springs and bogs—Summits, cliff ledges and screes

Of all the habitats which Scotland has to offer to the naturalist, the mountains and moorland hold the most appeal. The fascination of the mountains is compounded of several ingredients. There is solitude but seldom loneliness; there is the challenge to climp up to the high tops and always there is the lure of rare plant and animal life.

HEATHER MOORS

To write of one habitat is, however, misleading for there are many, but in a book of this modest size it is convenient to group them together. Between the floor of the glens and the upper mountain slopes is a wide altitudinal zone which over large areas is occupied by heather moor dominated by ling *Calluna vulgaris*. The flowering of the heather in August and September is one of the most splendid sights in Scotland. The geographical distribution is principally north of the Highland Boundary Fault, attaining the greatest density in the central and eastern Grampians. These moors are stocked with grouse and sheep, whose grazing helps to maintain the habitat which otherwise in most cases would develop into forest; but the status cannot be maintained by grazing alone, and periodically the moorland is subjected to burning known as the muir-burn. Whilst this practice improves the quality of the heather, it reduces the variety of plants within the community.

Over large areas of level ground and of gentle slopes, an almost pure community of ling is established above *Cladonia* lichens and mosses. In the drier areas, bell-heather *Erica cinerea* and the dwarf evergreen cowberry *Vaccinium vitis-idaea* may be present. Another evergreen shrub, bearberry *Arctostaphylos uva-ursi*, grows under heather in some places in the central and eastern Grampians. In this family also is the bilberry *Vaccinium myrtillus*, which is widely distributed over the central and northern Highlands, growing sometimes in association with crowberry *Empetrum* spp. Gorse *Ulex europaeus* is generally of very local occurrence on the moors, although it grows in profusion in certain areas in other habitats. The Welsh gorse *U. galli* and the dwarf gorse *U. minor* reach only southern Scotland, but petty whin *Genista anglica* is found mainly in the central and eastern Grampians and has decreased considerably in the Southern Uplands in the last forty years. Where drainage is poor, cross-leaved heath *Erica tetralix*, deer-grass *Trichophorum caespitosum*, heath lousewort *Pedicularis sylvatica* and heath rush *Juncus squarrosus* enter the plant community. Bracken *Pteridium aquilinum* grows in some parts in shaded or sheltered conditions below the 1,500ft level, and has spread in recent years due to the intensity of sheep-grazing. The Department of Agriculture has calculated that the area covered by this plant amounts to about 500,000 acres. Flowers which grow in association with heather include the heath spotted orchid *Dactylorchis maculata ericetorum*, bitter-vetch *Lathyrus montanus*, common milkwort *Polygala vulgaris* and dwarf cornel *Chamaepericlymenum succicum*. Nor must the harebell or Scottish bluebell *Campanula rotundifolia*, which grows on richer soil, be overlooked. An extremely rare plant, the blue heath *Phyllodoce caerulea*, grows in heather on the Sow of Atholl in Perthshire, the only known locality in Britain.

The common ground-hopper *Tetrix undulata* and the mottled grasshopper *Myrmeleottetix maculatus* are widely distributed on dry heather moor; in the moister parts the meadow-grasshopper *Chorthippus parallelus* is widespread, but the range of the bog bush-cricket *Metrioptera brachyptera* does not extend beyond southern Scotland. The moorland attracts a number of butterflies. The small heath *Coenonympha pamphilus* is the commonest species. The large heath *C. tullia* is essentially a northern butterfly, locally common over much of Scotland on boggy moorland. The form found in most of the country is the subspecies *scotica*, which is of pale colouration; but in parts of the Southern Uplands the typical form exists. Green hairstreaks *Callophrys rubi* occur in some places where there is gorse scrub, but this species is mainly restricted to the Central Lowlands. The marsh-fritillary *Euphydryas aurinia*, whose larvae feed on devil's bit, flies over damp meadows and boggy ground in June; the Irish subspecies *hibernica* has been recorded from the western coast of Argyll. A larger fritillary, which is more typical of heather moor, is the dark green *Argynnis aglaja*: it is locally widespread over the country. There remain the two butterflies which are of most interest to the lepidopterist: the Scotch Argus *Erebia aethiops* and the mountain ringlet *E. epiphron*. The former is much the more widespread and can be seen flying in sunshine over moorland, on the outskirts of plantations, in July and August. The caterpillars feed on the purple moor-grass. The mountain ringlet is a sun-lover, too, and occurs in scattered localities in the western Grampians, on the higher moorland and mountain slopes, at or above 1,500ft; the food plant of the caterpillars is mat-grass.

The heather moor provides a habitat for a large number of moths, several of which are of considerable size. The emperor *Saturnia pavonia* emerges in April and flies over the heather during the daytime. The northern subspecies of the oak eggar *Lasiocampa quercus callunae* appears in May and June; but the most abundant of these large insects is the fox moth *Macrothylacia rubi*, the males of which can be seen careering rapidly here and there over the heather in June. The larvae of many moths feed indiscriminately on various ericaceous shrubs, but the caterpillars of the northern spinach moth *Eulithis populata* are largely restricted to bilberry. This moth is a somewhat variable species: in the far north, for instance, its wings are almost black. Bearberry is the food plant of the caterpillars of the small dark yellow underwing *Anarta cordigera*; although an Arctic species, this moth frequents moorland below 1,000ft in parts of the Central Highlands. The larvae of a related species, the beautiful yellow underwing *A. myrtilli*, feed on heather; this moth is much more widespread. The little black-coloured chimney-sweeper *Odezia atrata*, whose larvae feed on pig nut, is a local species not particularly associated with moorland; but I have seen it flying in sunshine near its food plant on moorland above Loch Tay. Plants of the common milkwort may have on them in the late summer the caterpillars of the small purple barred *Phytometra viridaria*, which is well distributed over Scotland except in the Northern Highlands. Eyebright is the food of the two small moths, pretty pinion *Perizoma blandiata* and heath rivulet *P. minorata ericetata*, both of which are more plentiful in Scotland than elsewhere in Britain. On heath bedstraw may be seen the larvae of the small argent and sable *Epirrhoe tristata* and the purple bar *Cosmorhoe ocellata*. Both these species are more numerous in Scotland than elsewhere. In addition, all the common moorland species occur.

In summer, bumble-bees of various kinds move clumsily among the bells of heather. The small earth bumble-bee *Bombus lucorum* virtually replaces the large earth bumble-bee *B. terrestis*; and the small garden bumble-bee *B. hortorum* outnumbers the large garden bumble-bee *B. ruderatus*. Two more small bumble-bees are the early-nesting *B. pratorum*, which is plentiful; and the heath *B. jonellus*, which is more specifically associated with heathland than are some others, but which is only of local occurrence. Two large species are the stone bumble-bee *B. lapidarius*, which is recorded from various places in eastern Scotland, and the much more uncommon great yellow bumble-bee *B. distinguendus*, which is a northern species. The colouring of bumble-bees tends to be rather variable; and some species, of which the common carder *B. agrorum* is an example, have a different colour form in Scotland. The red ants *Myrmica* spp and the yellow hill ants *Acanthomyops flavus* are common inhabitants of the moors.

The unfortunate deer and sheep are troubled with certain species of parasitic flies. There are many flies in this habitat, but these must be left to the attention of the dedicated dipterist. Ground-beetles are common on moorland, extremes in size being represented by the large *Carabus clathratus*, a northern but local species, and by the common though much smaller *Bradycellus ruficollis*. Various plant-eating beetles are numerous, particularly weevils such as *Micrelus ericae* and *Lochmaea suturalis* on heather.

Although there is some overlap between birds of moor and mountain, a fairly distinct community of moorland breeding birds exists. Two species are pre-eminent on the moors in early summer: the meadow-pipit and the cuckoo. One cannot walk many paces on the Speyside moors in summer without seeing and hearing both. Whinchats are common summer visitors, frequenting grass-heath communities, and they can be seen on telephone wires in the glens and along the roadsides of the Sutherland moors. Stonechats also love perching-posts above the general level of vegetation; areas of gorse are a typical habitat, but rank heather comes a close second. The darker Hebridean form replaces the type in some western localities. Wheatears are by no means restricted to heather moors, but I have found them plentiful on moorland in south-east Sutherland and on the Stoer peninsula in the west. Linnets are widespread in Scotland, and apart from their favourite habitat of gorse are found perhaps more especially in heathery coastal areas. Twites are largely confined during the breeding season to the northern half of Britain. They resemble linnets, to which they are closely related; but they are distinctly darker and the pink rump of the male is diagnostic. The main stronghold is in Wester Ross and Sutherland, and there has been a contraction of range up to a few years ago. Breeding almost ceased south of the Highlands and the birds became very scarce in the north-east. A slow recovery of lost territory now appears to be taking place: by 1971 breeding was of regular occurrence in south-west Scotland and in 1973 I observed twites on the Banffshire moors. The yellowhammer is plentiful on moors wherever gorse and broom grow. Numbers of red grouse vary from estate to estate, and like other game-birds these fluctuate considerably from season to season; but the Perthshire moors around Aberfeldy are heavily stocked.

Most of the waders have their territory in the bogs and on the high ground; but mention can be made in this section of the golden plover: a familiar figure standing anxiously on the peat hummocks, the minor-keyed music of his plaintive call-note is completely in harmony with the mist and rain of these lonely wastes. Another nesting wader of this terrain, but with a much more restricted range, is the greenshank. The nests of this wader are often located near a piece of dead wood or a boulder; they are found sparingly in Inverness-shire and neighbouring counties and rather more plentifully in the Lewisian Gneiss country of Wester Ross and Sutherland.

Most of the predatory birds nest at sites away from the moor, but carrion and hooded crows and ravens spend much of their time there and occasionally nest among thick heather or low rocks. The carrion crow is the species in the Southern Uplands and Central Lowlands; and north of the Highland Boundary Fault there is an extensive zone where hybridisation between the hooded and the carrion crow occurs, but here the hoodie is the dominant species and in northern Scotland it reigns supreme. Recently, however, there have been reports of an increase of carrion crows in the Central Highlands at the expense of the hoodies. Kestrels, too, build their nests in a clump of heather, and they are frequently seen hovering over the moors. The smallest falcon, the merlin, makes its scrape among the heather; it is widely distributed on dry heather moors, but very local, and it has become scarcer in recent years although well established in Galloway. After the breeding season, these birds move to the coast, where they form small communal roosts. The hen-harrier is a bird of moorland valleys below 1,000ft; a fuller reference has been made in Chapter Seven both to this species and to the short-eared owl (see page 152). Like the merlins, hen-harriers have been observed in south-west Scotland, roosting communally in scrub near the shore during the wintertime. Montagu's harrier bred several times in Perthshire in the 1950s and at least once in Galloway, and it may yet do so again.

The largest mammal of the heather moor is the red deer. Red deer are well distributed in the hill country of the Grampians and the Northern Highlands, but are very scarce in the north-east lowlands. They are absent from the Central Lowlands and Southern Uplands, except for a few in Ayrshire and Kirkcudbrightshire. The principal food is ericaceous shrubs and grasses, consumed in the early morning and in the evening. Scottish red deer are less heavy and in general have smaller antlers than do their counterparts in continental woodlands, where the richer food-supply and greater shelter produce larger beasts. In some of the Scottish deer forests the animals are given supplementary food in the wintertime, and this has resulted in semi-tame animals which can be seen along the roadsides in some localities during the summer. The hinds in calf seek out a rocky cleft or a remote stretch of heather in June, and there the birth takes place. The stags are normally silent except during the autumnal rut, when the hills echo with their defiant challenge to rival stags.

Foxes are common on the moors, for there is food in plenty from ground-beetles to nesting birds and the carcasses of sheep. J. D. Lockie assessed the breeding density in Wester Ross at one pair on between 9,500 and 13,500 acres, and Dr A. Watson's assessment of the population on the deer forests of north-east Scotland was one pair of foxes per 3,200 acres. There is some divergence of opinion regarding the specific character of the Scottish foxes. It has been suggested that these hill foxes are grey of colour and belong to the Scandinavian subspecies, unlike the English ones which are of the Central European race. Lea MacNally, however, states from his wide experience that all of which he has had knowledge have been as the red English fox. Certainly some English foxes have a distinctly grey coat, but this may be an indication of age. Clearly there is scope for more research on this subject. The wild cat is another mammalian predator which frequents moorland, but its den is likely to be on steep rocky crags. This animal is definitely on the increase. D. Jenkins in 1962 recorded the greatest numbers in Inverness-shire and Aberdeenshire, with an increase proceeding in certain neighbouring counties. He stated that numbers were low in the north and the west, but it is probable that some increase has taken place since then, for in various localities in these regions today the wild cat is considered to be relatively common.

The blue mountain hare is typically an animal of the higher moors between 1,000 and 2,500ft, although it occurs occasionally above and below these levels. The greatest densities are found on moorland in the county of Roxburghshire, on the grouse moors of Perthshire and in the east-coast counties from Kincardineshire to Nairnshire. The Southern Uplands stock resulted from introductions from 1834 onwards. The term 'blue' is misleading, since the summer coat is greyish-brown, and at a distance the animal is not easy to distinguish from a brown hare except by its smaller size. Small mammals such as shrews, voles and fieldmice are common inhabitants of the moors. Young frogs are conspicuous everywhere on the moorland in early summer and toads are common. Lizards are ubiquitous among the heather, but adders and slow worms, although widespread, vary in abundance.

SPRINGS AND BOGS

In the wetter western parts of Scotland there are large areas of blanket bog. From the standpoint of the artist or photographer who isolates a portion of the whole to create a miniature of beauty, the bogs can be not unattractive. Sunlight glinting on a boggy pool surrounded by the vivid emerald cushions of Sphagnum moss, white plumes of cotton-grass limply hanging their heads, the bottle-green neck of a mallard rising out of a tussock of moor grass—these are cameos etched on a broad landscape which, admittedly, can be forbidding in its drear monotones. For the plant ecologist too, there is interest in

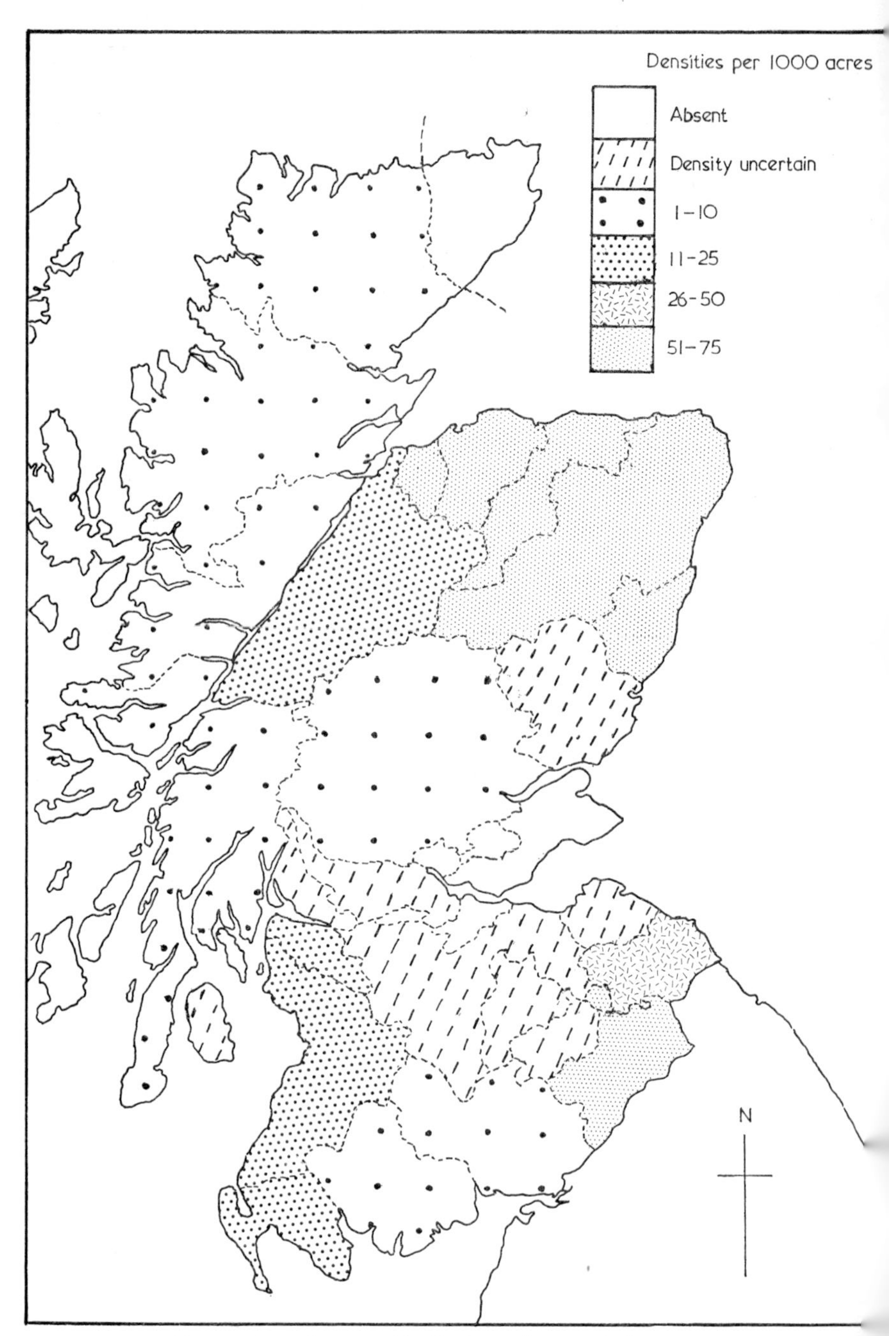

Fig 5 Map showing distribution of mountain hares in Scotland

differentiating the various plant associations, but to the rambler who tries to negotiate his weary way from hummock to hummock, with many a reluctant detour, the traversing of a blanket bog is an exhausting process. These bogs are not confined to the west: the traveller from Lairg, northwards through Altnaharra to Tongue, will see a part of the vast bogs of central and eastern Sutherland. In lowering cloud and rain these bleak wastes are unequalled in their utter desolation.

The characteristic plants are ling, cross-leaved heath, deer-grass, purple moor-grass and bog-myrtle *Myrica gale*, with cotton-grass *Eriophorum* spp, bog-asphodel *Narthecium ossifragum*, sundew *Drosera rotundifolia*, butterwort *Pinguicula vulgaris* and *Sphagnum* spp in the wettest parts. Crowberry and bog-bilberry *Vaccinium uliginosum* are other constituents in various places, more especially in eastern areas. A very rare rush-like plant *Scheuchzeria palustris* grows among deer-grass and cotton-grass on Rannoch Moor, its only known station in Scotland. Cranberry *Vaccinium oxycoccus* and small cranberry *V. microcarpum* are local plants of boggy areas, the latter restricted to the Grampians.

Where there is movement of water in the soil, marsh vegetation develops in which sedges and rushes of various kinds predominate. There are small areas of open water in some places, and a swamp flora builds up round the margins. This type of habitat in Scotland is known as a moss and is largely confined to the southern part of the country. Bemerside Moss in Berwickshire can be cited as an example and here are found such typically marginal aquatic plants as bur-marigold *Bidens tripartita*, marsh-cinquefoil *Potentilla palustris*, celery-leaved crowfoot *Ranunculus sceleratus* and broad-leaved reed-mace *Typha latifolia*. In the limited areas where the water derives from basic rocks, certain distinctive species occur. The black bog-rush *Schoenus nigricans* is one such. This forms almost pure communities in a restricted number of localities along the western seaboard. A number of sedges, although not calcicoles, nevertheless have a tolerance which enables them to grow in calcareous soil. Marsh arrow-grass is a local indicator of somewhat calcareous conditions; and the much more uncommon fen-sedge *Cladium mariscus*, which grows in some western districts, signifies strongly basic conditions.

An important wetland habitat in the mountains is that of the many springs which issue from the mountainside, carrying various mineral salts to a bordering 'flushed' area. A colourful assortment of flowers is present in this situation. These include chickweed willow-herb *Epilobium alsinifolium* and alpine willow-herb *E. anagallidifolium*, with their drooping pink flowers; the yellow mountain saxifrage *Saxifraga aizoides*; the starry saxifrage *S. stellaris*, whose white petals provide a perfect foil for the orange-red stamens; the Scottish asphodel *Tofieldia pusilla*, with its cool green-tinged white blossoms; and the little alpine meadow-rue *Thalictrum alpinum*, elegant in flower and foliage. The butterworts *Pinguicula vulgaris* and *P. lusitanica* like a moist situation and thrive as well in the flushes as in the bogs, though the last-named plant is confined to the western part of the country. A characteristic feature of these mountain springs is the dense carpet of mosses such as *Philonotis fontana* and *Cratoneuron commutatum*, in colours ranging from golden and glaucous greens to orange-brown.

The aquatic habitats attract a number of moths. Cotton-grass is the food of the larvae of Haworth's minor *Celaena haworthii*, a small moth which is more common in Scotland than elsewhere in Britain. The larvae of another small northern species, the Manchester treble-bar *Carsia sororiata*, feed on cranberry and cowberry. In the autumn and early spring the caterpillars of the yellow ringed carpet *Entephria flavicinctata* can be seen on the leaves of the yellow mountain saxifrage. Bog-myrtle is a favourite food plant of several species. The ringed carpet *Cleora cinctaria*, which is locally abundant in the southern Grampians, has a separate subspecies in Scotland; in the British Isles generally the larvae feed on various plants, but in Scotland bog-myrtle is reported to be the principal food. A Scottish speciality, the Rannoch brindled beauty *Poecilopsis lapponaria* has a very local distribution in Perthshire, Inverness-shire and Ross-shire, and other bog-myrtle feeders include the larvae of the red sword-grass *Xylena vetusta*, the sweet-gale moth *Acronyeta euphorbiae* and the light knot-grass *A. menyanthidis*: all are northern species. Of the bumble-bees, the moss carder *Bombus muscorum* frequents boggy heaths. A small plant-bug *Teloleuca pellucens*, a species of northern Europe, is found among Sphagnum in the hill bogs of Perthshire and Inverness-shire.

Mallard and teal nest in some of the bogs and the gulleries of black-headed and common gulls are sited on a number of the mosses. Waders such as snipe, dunlin and curlew are widespread breeders. A rare nesting species of marshy ground is the wood sandpiper, which has become a regular breeding bird in very small numbers in Sutherland and probably also in the Grampians. Its close relative, the green sandpiper, has nested in recent years and it too may become a regular nester.

SUMMITS, CLIFF LEDGES AND SCREES

Of all the various plant communities which exist in Scotland, it is the alpines on the mountains which draw the wild-flower lover like a magnet; and the fascination which they hold for their devotees can be understood readily by anyone who has ever seen the tight pink-studded cushions of the moss-campion in a crevice, the creamy white flowers of mountain avens cascading over the rocks or the purple saxifrage staining the April snows. These montane plants occupy a variety of situations: on the summit plateaux, in the high corries, on the cliff ledges and in the interstices of the screes. A number of these alpines are quite rare and, it is hardly necessary to add, deserve to be left alone rather than to be uprooted to wither in totally uncongenial conditions in an urban rockery. It is ironic that while egg-collecting has been illegal for many years, at the time of writing, an Act to protect wild plants is still awaited.

Mountain flowers can be separated into two main categories: those associated with acid rocks and those for which a calcareous soil is essential. In the former category there are plants with a wide degree of tolerance which grow on a variety of rock types. In the latter are found most of the rare and beautiful alpines. Granite is an acid rock and the Cairngorms are one of the largest granite masses in the British Isles. The summit plateau which is the highest level ground in Britain is easily, some say too easily,

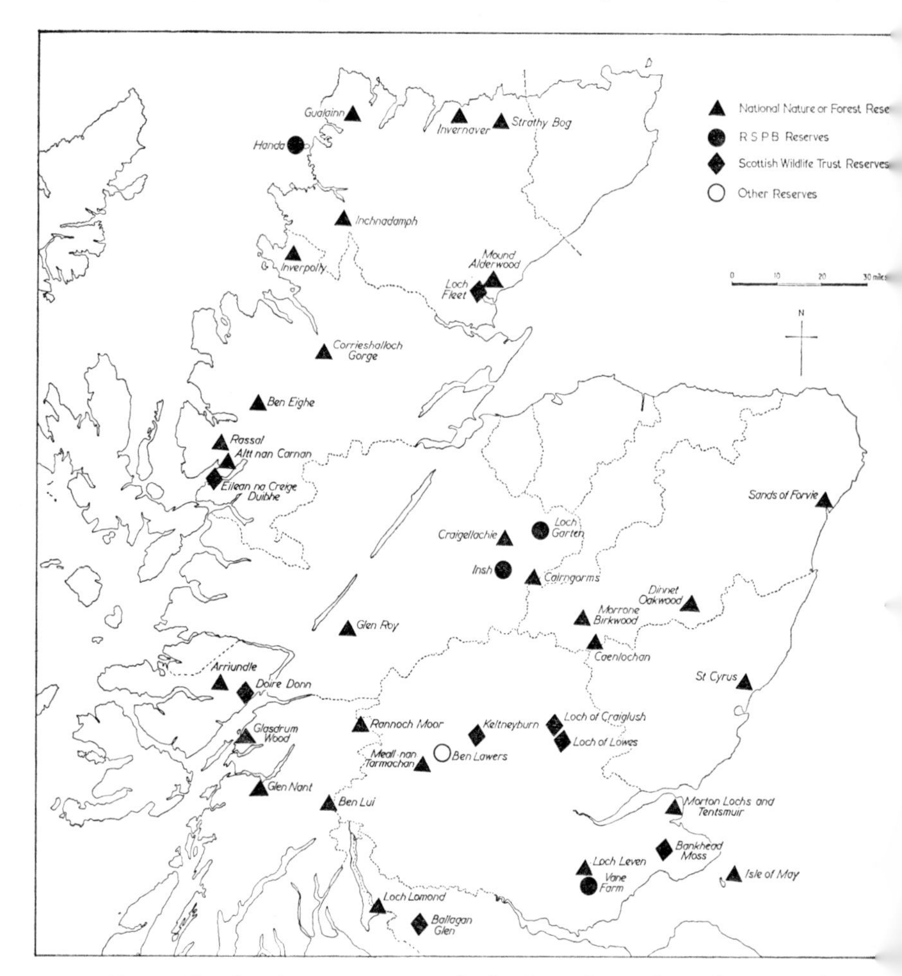

Fig 6 Map showing nature reserves in the Grampians and North-west Highlands

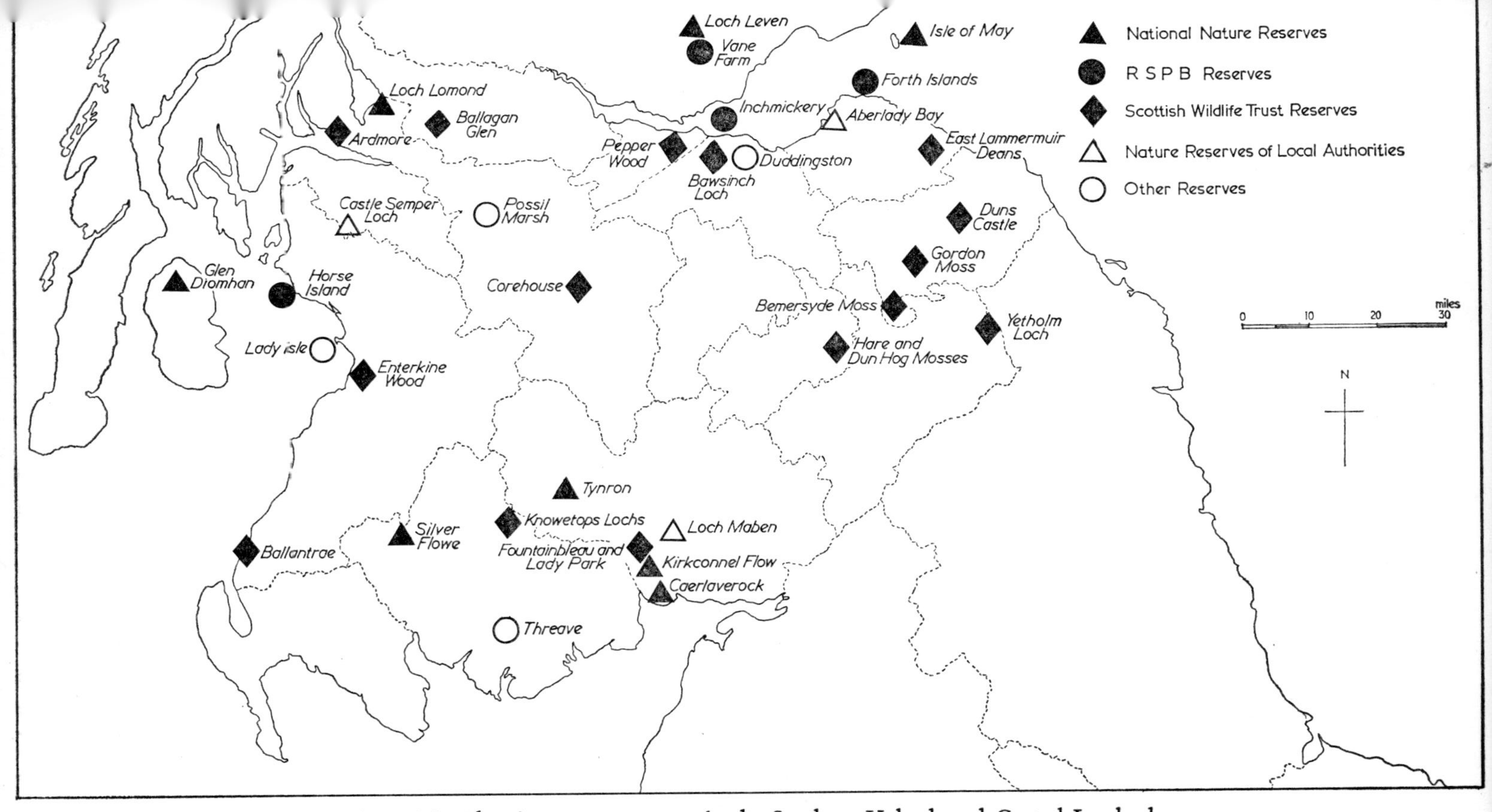

Fig 7 Map showing nature reserves in the Southern Uplands and Central Lowlands

reached via the chairlift. Be that as it may, there is a vast area in which the hill walker will meet few, if any, other people. This plateau is an exhilarating place on a fine summer's day. The tiredness arising from the uphill walk soon vanishes as the lungs absorb the clear mountain air. The dominant vegetation is the woolly fringe moss *Rhacomitrium lanuginosum*, which grows both on the peaty soil and on the rocks. This is the food of the red deer in the summer when they move up to the tops to escape the unwelcome attentions of the flies. The exposed wind-swept nature of these heights prevents many other plants from living there. There are crowberry and dwarf willow *Salix herbacea* which may fairly be regarded as the smallest tree in Britain. One day on this plateau I came across the tiny gnarled trunk of a dwarf willow only an inch or so high, long bent over by the wind and looking for all the world like a prize specimen of the Japanese art of Bonzai. Two drab-looking plants common in this habitat are the stiff sedge *Carex bigelowii* and the three-leaved rush *Juncus trifidus*, as well as a speciality of the Cairngorms: the uncommon curved wood-rush *Luzula arcuata*. A splash of colour is added by the pink flowers of the alpine azalea *Loiseleuria procumbens*, which drapes itself over the rocks.

Plants on rock ledges and in crevices on the Cairngorms include the northern rockcress *Cardaminopsis petraea*, mountain speedwell *Veronica alpina*, starry saxifrage, moss-campion, alpine bistort *Polygonum viviparum*, alpine saw-wort *Saussurea alpina* and, especially, mountain hawkweeds. There are some 260 species of hawkweed, of which about one-third can be described as mountain hawkweeds, but only the specialist is likely to concern himself with separating the species. In the corries grow alpine lady's mantle *Alchemilla alpina*, the sedge *Carex lachenalii*, the rare brook saxifrage *Saxifraga rivularis* and various mosses and liverworts.

Separated from the main granite mass of the Cairngorms is the mountain outpost of Lochnagar on the Balmoral estate. In its high corries grow an assemblage of plants of surprising variety and rarity for so acid a rock. These plants include the very rare grass *Poa flexuosa* and its equally rare hybrid *P. jentlandica* which grow on barren stony slopes, the highland cudweed *Gnaphalium norvegicum* and the spectacular blue sowthistle *Cicerbita alpina*, as well as several rare sedges.

Granite of course is by no means the only acid rock type, and much farther north we can look at the flora of the mainly acid rocks of Torridonian Sandstone, Lewisian Gneiss and Cambrian Basal Quartzite in the north-west. A plant which is particularly plentiful on rocky outcrops and also in bogs is the alpine bearberry *Arctous alpina*. Another shrub typical of this region is the dwarf juniper *Juniperus communis nana*, which grows on screes and morainic deposits. Screes have been formed in a number of places, notably on Cambrian Quartzite. Where there is some safeguard from frost, either by snow cover or by proximity to the coastline, the parsley fern *Cryptogramma crispa* is the characteristic coloniser of screes together with mosses, tufted hairgrass *Deschampsia caespitosa* and, at a rather later stage, ling. The dwarf cornel *Chamaepericlymenum suecicum* grows in montane heaths, although not so plentifully as in the Grampians. On high ground the flowers of cloudberry *Rubus chamaemorus* can be seen, but seldom the amber fruit, for many are male plants and those that are female are, for some reason, reluctant to fruit. The alpine lady-fern *Athyrium alpestre* grows sometimes abundantly on shady rock ledges up to 3,600ft, and clumps of mountain buckler fern *Thelypteris limbo-sperma* with their strong scent of lemons thrive from the roadsides to far up the mountain slopes. The Norwegian mugwort *Artemisia norvegica* was discovered in Britain only as recently as 1950, on Torridonian Sandstone in Wester Ross; it is now known to occur on two hills and is common in the habitat in which it is found.

It is, however, to the limited areas of either calcareous rock or that containing other basic minerals, that we must go if we want to see the richest variety of beautiful alpines. These comprise three principal areas. The first area is a belt of calcareous schists within the main Dalradian mass, stretching diagonally north-eastwards from Ben Lui on the Argyllshire–Perthshire boundary through Meall nan Tarmachan, Ben Lawers, Ben Vrackie and Ben-a-Ghlo to end in a blaze of glory at Caenlochan Glen in Angus, where there are more alpine species, some of great rarity, than anywhere else in Britain. The second area is much farther north, a narrow belt of Cambrian limestone stretching from Ullapool in Ross-shire northwards through Knockan, Elphin and Inchnadamph to Durness on the north Sutherland coast. The third area consists of the basaltic lavas of the Ardnamurchan peninsula and Morven in the west. Much smaller localities, where basic rock produces a rich flora, include a belt of hornblende schist on the slopes of Ben Hope in Sutherland and Silurian shales in the vicinity of the Moffat Water in the Southern Uplands.

Of all these localities I suppose it is true to say that Ben Lawers has held the greatest attraction for botanists over the longest period of time, that is, for over 180 years. This mountain is deservedly famous, but there are other areas almost as rich and one suspects that its comparative accessibility has something to do with its popularity. The flora of Ben Lawers includes not only lime-loving plants but other montane species which are indifferent to soil type. The high-level ledges just below the 3,984ft summit are natural rock-gardens which are full of colour in midsummer, continually kept moist by water dripping from the cliff face above and constantly replenishing their mineral content. Such a ledge may have moss-campion and a similar but less colourful cushion plant, the mossy cyphel *Cherleria sedoides*, mountain sorrel *Oxyria digyna*, globe-flower *Trollius europaeus*, mossy saxifrage *Saxifraga hypnoides*, alpine willow-herb *Epilobium anagallidifolium*, red campion *Silene dioica*, alpine scurvy-grass *Cochlearia alpina* and green spleenwort *Asplenium viride*: a miscellaneous collection of widely differing plants. Among the treasures of Ben Lawers are the purplish-pink flowers of boreal fleabane *Erigeron borealis*, rock whitlow-grass *Draba norvegica*, several blue-flowered alpines—alpine forget-me-not *Myosotis alpestris*, rock speedwell *Veronica fruticans* and the snow gentian *Gentiana nivalis*—and above all the drooping saxifrage *Saxifraga cernua.* On the opposite side of the Tay–Glen Lyon road, at Lochan na Lairige, there are crags bearing a rich flora: I have found there the beautiful orange-yellow flowers of the alpine cinquefoil *Potentilla crantzii* which also grows on Ben Lawers itself. By the concrete dam at this spot, two plants new to Britain were discovered on 9 August 1964. One was an arctic-alpine moss *Aongstroemia longipes*, dominant over several acres, and the other was a European liverwort *Fossombronia incurva.* It is interesting to speculate how these plants arrived. In view of the fact that this is a much-frequented locality, which at times positively swarms with botanists, it is likely that the plants are recent arrivals. Initially one would suspect some connection with material brought in by the contractors, but in the case of the moss, at least, this does not seem a likely possibility.

At the head of the two Angus glens of Isla and Clova is a large area of nearly 9,000 acres, where rich vegetation grows on high ledges beyond the reach of animals, domestic or wild. Here in Glen Doll where the eagle soars, or on the slopes of Glen Fee where the red hinds graze the mountain pasture, or in remote Caenlochan Glen, the lover of wild flowers can find, as well as many of those plants listed in the previous paragraph, the dwarf mountain forms of golden rod *Solidago virgaurea* and sea-pink *Armeria maritima*, alpine lady's mantle *Alchemilla alpina* and its much rarer relation, *A. conjuncta*, which occurs in a native state elsewhere in Britain only in Arran, northern bedstraw *Gallium boreale*, *Sibbaldia procumbens*, the rare alpine saxifrage *Saxifraga nivalis*, alpine milk-vetch *Astragalus alpinus*, blue sow-thistle *Cicerbita alpina* and alpine catchfly *Lychnis alpina.* The last-named is a plant which is often associated with a high metalliferous content in the underlying rocks. Another rare member of this genus, the red catchfly *L. viscaria*, also grows in Angus and farther south on the volcanic rocks of Arthur's Seat at Edinburgh.

The mountain avens *Dryas octopetala* occurs in Angus, but it is more abundant farther north. It is the most conspicuous alpine of the Sutherland calcareous strata, where it trails over the rocks, with the slightest breeze revealing the silvery white underside of the leaves. The dark red helleborine *Epipactis atrorubens*, the Norwegian sandwort *Arenaria norvegica*, the whortle-leaved willow *Salix myrsinites*, the rare grass Don's twitch *Agropyron donianum* and the holly-fern *Polystichum lonchitis* are notable members of the flora on the Cambrian limestone of the Inchnadamph reserve. The grass derives its name from the Angus nurseryman George Don, who in 1810 discovered it growing on Ben Lawers; but it was not until 1951 that it was identified at Inchnadamph. This is just one indication of the possibilities which still exist of making exciting botanical discoveries on the Scottish mountains. A more dramatic example is that of the finding in the same year of an alpine not known to occur in the British Isles—not an isolated plant at that, but one carpeting the summit of an Inverness-shire hill. The flower in question is *Diapensia lapponica*, a white-flowered shrubby plant of the Arctic, and the finder was a visiting English naturalist.

Some of the moths which fly at the higher levels have already been mentioned in the section on heather moors (see page 170), and it remains but to refer to the more specifically montane element. The black mountain moth *Psodos coracina* flies in July sunshine and occurs even as high as the Cairngorm plateau, in such numbers that Seton Gordon states that these moths are an attraction to the black-headed gulls which have been a frequent sight over the Cairngorm summits in recent summers. The netted mountain moth *Semiothisa carbonaria* can be seen high in the Grampians, flying somewhat earlier than the preceding species, in the spring sunshine. The shrubs of the *Vaccinium* genus are the food of the larvae of the broad-bordered white underwing *Anarta melanopa*, an Arctic species. The caterpillars of the northern dart *Xestia alpicola* also feed on these shrubs, although perhaps more commonly on crowberry. The forewings of this moth have an attractive pebbly appearance. The grey mountain carpet moth *Entephria caesiata* has an even more striking cryptic colouration which perfectly matches the lichen on the rocks on which it rests. A very rare moth of the mountains is the mountain burnet *Zygaena exulans*, known only from a locality in the eastern part of the Central Highlands. It seems that almost all the real mountain species of moth are day-flying, and the mountain burnet is no exception. Perhaps this is because there is less danger of predation and therefore no need for nocturnal habits.

There is one of the bumble-bees which is more characteristic of montane situations than the others: this is the bilberry or mountain bumble-bee *Bombus lapponicus*, a medium-sized, red-tailed bumble-bee of local distribution in the Highlands.

The Scottish mountains have a distinctive spider fauna. A few spiders are typical of the screes and include a wolf-spider *Lycosa traillii* and the tiny *Theridion bellicosum*. More species live under the stones among the summit clitter. The very rare wolf-spider *Arctosa alpigena*, large and conspicuously marked, is a high-altitude spider and in Britain is known from the Cairngorms. Most species belong to the great family Linyphiidae, the familiar small 'money' spiders of childhood. They include *Entelecara errata*, *Erigone tirolensis*, *Tiso aestivus* and *Collinsia holmgreni*. The last-named occurs both in the Grampians and in the Northern Highlands and is reported to be abundant on the summit of Braeriach in the Cairngorms. Perhaps the rarest species in this habitat is the Linyphid *Macrargus rufus carpenteri* which has been recorded from Ben Macdhui in the Cairngorms. Only a few individuals have ever been observed in Britain: in the Lake District and in Scotland. It is a small montane subspecies of a common woodland spider.

One would not expect to find many slugs and snails in the exposed and drying conditions of the mountain summits, since both kinds require humidity; but a few species, including the common snail of gardens, *Helix hortensis*, are able to live above the 3,000ft level in moist sheltered conditions.

Birds of high altitude are few in species and thin on the ground. Of the small passerines, scattered pairs of meadow-pipits and wheatears hold summer territories on high ground from the Southern Uplands to the Cairngorms, but they come down to lower ground after breeding. In contrast, the snow-bunting is essentially a mountain bird where breeding is concerned, seldom nesting below 3,500ft. Even in the winter, when it is more widely dispersed, numbers frequent the high tops, remaining until late in the spring. Nearly all these winter birds are seasonal migrants, since very few pairs nest in Scotland. Numbers of both winter visitors and summer breeders fluctuate from year to year, but it appears that nesting is becoming more regular. The birds are elusive nesters in the remote high corries of the Central Highlands, their twittering call-notes carried on the strong winds so frequent at this altitude. A winter flock in undulating flight is a beautiful sight and inspired the seventeenth-century naturalist Sir Robert Sibbald to use the name 'snow-fleck'. The ring ouzel is the upland blackbird, replacing that species at about the 1,000ft level and haunting the higher glens, rocky slopes and screes mainly up to 2,000ft, although birds are occasionally seen above this level. It has become considerably scarcer in recent years, and it has been suggested that, at the lower levels of its range, the aggressive competition of its relation prevents successful breeding. The thrice-repeated piping call, whose wildness matches the scenery, appears to possess a ventriloquial quality; at any rate, the source of the call can sometimes be difficult to locate. Though scarce, ring ouzels are well distributed from the cleughs of the Southern Uplands to the Sutherland hillsides.

Three waders have their nesting haunts on the high tops. The dunlin is by no means restricted to high ground but nests on boggy ground and peat-hags from sea-level to the Cairngorm plateau. The breeding plumage with black belly patch is handsome, contrasted with the plain colours donned for the winter mud-flats. The golden plover also nests on high ground as well as on moorland at much lower levels. The third and rarest wader is the dotterel, a bird whose trusting tameness is an endearing trait. It is closely related to the golden plover, but there are differences of behaviour. The altitudinal zone of the golden plover is from 500ft to just over 3,000ft, whereas that of the dotterel is 3,000ft upwards. The golden plover is handsome, even garishly so; but the dotterel is elegant with a bold eyestripe, mushroom brown upper breast and chestnut lower breast separated by a white pectoral band above a black belly. Such a description may make the bird appear gaudy, but it is far from that; and it is astonishing how well camouflaged the dotterel really is in its mountain habitat. Another difference which has been mentioned earlier is that, while the golden plover on its nesting territory tends to be wary and suspicious, the dotterel is often excessively tame, so much so that the Gaelic name for it means 'the fool of the moss'. Yet another dissimilarity lies in the fact that the male bird performs the maternal duties of incubating the eggs and looking after the young. This it has sometimes to do in very inclement conditions, for snow falls in early summer and icy blasts sweep across the plateaux. Small numbers breed in the Grampians and in Ross-shire and in the last year or so there has been a slight improvement in status.

The game-bird of the high tops is the ptarmigan, usually found in the Grampians only between 2,500 and 4,000ft, although I have more than once come upon pairs and females with chicks below the lower level in the Lairig Ghru pass. They become scarce at the upper level and have the greatest numbers concentrated between 3,000 and 3,500ft. In the Western Highlands they breed at a rather lower altitude and in Sutherland, where they are far from common, they nest at lower levels still. The typical habitat of the ptarmigan is the boulder-strewn, lichen and moss-covered ground of the summits and corries; and here the guttural rattle is a familiar sound. It is a handsome bird in its grey and white plumage with a red wattle over the eye. The bird is unusual in having three moults every year. The autumn plumage is a transitional one, when the summer markings become grey and finer; and in its almost entirely white winter plumage the bird matches its snowy surroundings as perfectly as when the grey mottling of its summer plumage blends with the lichen-bedecked rocks. Each of these plumages must have a considerable survival value for the ptarmigan as the eagle flies overhead searching for prey.

Scottish mountains and golden eagles are bracketed together in the mind of the birdwatcher as are ham and eggs in the mind of the famished hill walker. The eagle breeds in the hilly areas of the western and central parts of the country, from Kirkcudbrightshire in the south, where a small population re-established itself in 1948, to Sutherland in the north. It has been estimated that there are between 110 and 130 pairs on the Scottish mainland, and the number is now probably rather higher than this. Each pair requires a minimum territory of 10,000 acres. From 1961 onwards there was a marked decline in the number of young reared successfully, and suspicion fell on the use of chlorinated hydrocarbons in insecticides at the sheep-dips. In 1966, the Government imposed a ban on the use of dieldrin for sheep-dips and since then sample surveys have indicated that the eagles' reproduction rate has returned to normal. The prey is varied, depending to some extent on locality, availability of prey and individual preferences. Frequently taken prey are red grouse, ptarmigan, mountain hare and carrion. Eagles are sometimes persecuted because lambs are killed occasionally; but although lambs and deer calves are sometimes taken, they form only a small proportion of the total food. Where the staple food is scarce, eagles will take a variety of birds, rabbits and rats. There are usually two eggs, and since incubation is begun with the laying of the first egg, the eaglets are uneven-aged. An unfortunate result of this is that in many cases the older eaglet becomes aggressive and kills the younger. The nest site, which often has a rock overhang, is by no means always situated at high altitudes: I have seen an eyrie as low as 300ft above sea-level near the west coast.

The buzzard is a fairly common bird in most of Scotland and appears to have fully recovered from the set-back caused by myxomatosis in the rabbit. At a distance it is sometimes mistaken for an eagle; but I feel that if there is uncertainty about identification, then the bird is almost sure to be a buzzard, for the eagle's powerful wings leave little room for doubt. The peregrine falcon, like the eagle, suffered a serious population decline from the late 1950s to 1963, but since then it has made a limited recovery. Just how limited is this recovery was revealed by D. A. Ratcliffe's survey in 1971. This showed that of the total territories surveyed in Britain, the occupation was only 54 per cent of the average for the 1930s. The position in Scotland, however, is much better and the occupancy percentage is 72 per cent. The improvement which has taken place has been largely in the inland areas of the Southern Uplands and the Central Highlands, where most known territories were occupied in 1971. The lack of improvement on the coastal cliffs is presumably due to marine pollution. The peregrine is a magnificent bird, handsome of plumage and with complete mastery of the air. Its clamant calls cannot fail to excite the bird lover. It would seem that, in the long term, the survival of this species will depend on effective control of pesticides. The raven builds his large nest on a cliff ledge usually below 2,000ft—often like that of the eagle's, with an overhang. Breeding is early and the first eggs are laid usually by the beginning of March. The snowy owl is visiting the Highlands rather more frequently, and a bird has summered on the Cairngorms for successive years. Suitable breeding habitats exist and it is possible that in due course the mainland, like Shetland, will have nesting snowy owls.

Red stags move to the high ground during the summer, probably in the main to escape the attention of troublesome flies. A large herd of several hundred on the Moine Mhor in the Cairngorms is a magnificent sight. The hinds, too, move to higher levels but keep separate from the stags. One summer day in Sutherland, I was watching a herd of thirty-two hinds on the summit skyline of Creachan Thormaid, when suddenly two of them had a disagreement. After a ceremonial bowing of heads and touching noses—traditional preliminaries in the manner of duellists—they simultaneously reared up on their hindlegs and boxed with their forefeet like jack-hares. Stags will also fight in this way during the summer, although when the rut is under way antlers are brought into play. The only other species of deer to be seen at high levels is the introduced herd of reindeer on

Sron an Aonaich, the northern outpost of Cairngorm. In scattered localities throughout the country, herds of wild goats roam the crags, but they are not always to be found at the highest levels. They are difficult to approach and have an uncanny ability to disappear without trace. Some of the billies grow magnificent horns and domestic goats which escape on to the hills revert rapidly to the wild type.

The mountain hare is a mammal which, although principally found on the high moorland, occurs also on the summit plateaux. It is specially adapted to its all-the-year-round existence on high ground for, like the ptarmigan, it changes to white in winter. It is rare for one to become completely white: the ears retain their normal colour and there is usually some trace of darker fur on the face. This is better cryptic colouration than pure white would be, since it blends with the rocky outcrops behind which the hare is forced to shelter in the frequent winter gales which sweep across the high tops. An alternative hideout is a tunnel under the snow, but even with these two kinds of shelter only a very tough creature could survive the extreme conditions in winter on the Scottish mountains. Small mammals, more especially shrews, are occasionally seen on high ground, particularly during the summer.

Mammalian predators are few at these heights. Foxes are opportunists, and although their earths are not usually located above 3,000ft, they are quite capable of hunting for prey—ptarmigan or mountain hare—on the summit plateaux. The wild cat also ranges over a wide altitudinal zone from sea-level, where I have known of it attacking domestic ducks, through glen and forest to high on the mountainside; but its typical breeding habitat is a boulder-strewn hillside or a well-vegetated rocky cliff. Unlike the domestic cat, the wild animal does not cover its droppings, so that the presence of these with their pungent odour indicates the animal's proximity. They are largely nocturnal in habit, sleeping in a rocky lair or dozing in the sunshine on a slab of rock.

It is fitting to conclude both this chapter and the book itself with a brief account of a rare and seldom seen mammal which, though in the British Isles not entirely restricted to Scotland, is nevertheless most numerous there. I refer to the pine marten. This fine animal belongs to the weasel family, and with its rich brown back fur, creamy white throat-patch and large, intelligent, appealing black eyes, it is an altogether attractive creature. Originally a woodland animal and still occurring in both the old Caledonian forests and in coniferous plantations, its stronghold in Scotland is the rocky ground of Wester Ross and Sutherland. The marten, like the wild cat, is nocturnal and even more difficult to observe; but in areas where they are relatively plentiful, they will sometimes turn up at garden bird-tables or even take food scraps from windows. There is little doubt that the numbers have increased considerably since the low point of near-extinction was reached at the beginning of the present century. The marten's safety was ensured by the practical concern of one or two enlightened landowners. Today many more people are anxious to protect not only the pine marten but the whole of Scotland's priceless heritage of wildlife. This, too, is cause for rejoicing.

Appendix

Areas of natural-history interest

THIS LIST IS of necessity a very selective one due to the immense area involved. Some localities mentioned in detail in the text are not listed again. It should be noted that for many of the reserves a permit is required for entry. The listing of a locality does not necessarily indicate a right of access; and although large parts of the Highlands are open to visitors during most of the year, some restriction of access is made during the shooting season. Additional localities are listed in the section on nature trails.

List of abbreviations

DCC	Dumfriesshire County Council
ELCC	East Lothian County Council
FC	Forestry Commission
FNR	Forest Nature Reserve
HBF	Highland Boundary Fault
LNR	Local Nature Reserve
MCC	Midlothian County Council
NNR	National Nature Reserve
NTS	National Trust for Scotland
RCC	Renfrewshire County Council
RSPB	Royal Society for the Protection of Birds
SSPWB	Scottish Society for the Protection of Wild Birds
SSSI	Site of Special Scientific Interest
SWT	Scottish Wildlife Trust
WT	Wildfowl Trust

Aberdeenshire

Bullers of Buchan, south of Peterhead. NK 1138. Large breeding colonies of sea-birds.
Dinnet Oakwood, Dee Valley. NNR. Permit required to visit.
Feugh Falls, near Banchory. NO 7095. Salmon leap.

Huntly, centred on NJ 5340. Area of geological interest with various igneous rocks.
Lochnagar, mountain south-east of Braemar. NO 2585. Rare alpines.
Loch of Skene, west of Aberdeen. NJ 7807. Wintering duck.
Loch of Strathbeg, south-east of Fraserburgh. NK 0759. Coastal loch of ornithological interest at all seasons.
Mar Forest, west of Braemar. Old Caledonian forest remnant with uncommon plants, red and roe deer, red squirrels, capercaillie, black game and crossbill.
Peterhead Bay, NK 1245. Wintering sea-duck.
Sands of Forvie, Newburgh, north of Aberdeen. NK 0026. NNR, 1,774 acres. Terns, eider, shelduck, wintering geese and duck.

Angus

Arbroath Cliffs, NO 6641. Rock formations, cliff plants, breeding sea-birds.
Balgavie Loch, east of Forfar. NO 5351. Great-crested grebes, wintering wildfowl.
Buddon Ness, northern tip of Tay estuary. NO 5530. Good viewpoint for wintering sea-duck. West of the Ness towards Monifieth, terns and little gulls.
Caenlochan, south of Braemar. NNR, 8,991 acres. Wide range of vegetation types and exceptionally rich alpine flora; wild cat, mountain hare, red and roe deer; varied bird life including five species of birds of prey. Seasonal restriction of access.
Glen Clova. May be taken as typical of Angus glens. Metamorphic rocks exhibit mineral zonation; varied bird life. At head of glen, red squirrel, wild cat, ring ouzel, golden eagle.
Invergowrie Bay, Tay estuary. NO 3529. Waders and wildfowl.
Loch of Forfar, west of Forfar. NO 4450. Wintering duck and whooper swans.
Loch of Lintrathen, north-east of Blairgowrie. NO 2754. Wintering duck, greylag geese, whooper swans.
Montrose Basin, NO 6857. Waders including less common passage migrants, wintering duck, pink-footed and greylag geese.
North Esk Estuary, south of St Cyrus. NO 7462. Waders and wildfowl.
Rescobie Loch, east of Forfar. NO 5151. Wintering wildfowl.

Argyllshire

Ardgour, region of north Argyll. Relict-valley pine woods, red and roe deer, wild goats, badgers, wild cats, golden eagle, black grouse, crossbills. Rare mineral strontianite from Strontian lead-mines.
Ardnamurchan, peninsular region of north-west Argyll. Geological interest with Tertiary ring dykes and Jurassic fossils at Kilchoan; rich flora on coastal grasslands, eg at Kilmory; alpines include northern rockcress on Ben Hiant.

Argyll National Forest Park, south-east Argyll. Wide range of vegetation types; otter, wild cat, roe deer; ptarmigan and birds of prey including golden eagle.
Arriundle Oakwood, NM 8464. FNR, 288 acres. Semi-natural woodland with rich bryophytic flora; wild cat; typical woodland birds including redstart.
Black Mount, mountainous area of north-east Argyll. Corrie Ba is largest corrie in Scotland; some relict pine woods; red and roe deer, wild cat; golden eagle, ptarmigan, black game, occasional greenshank.
Davaar Island, south-east Kintyre. NR 7620. Caves; wild goats; shelduck, terns.
Doire Donn, on Loch Linnhe. NN 0469. LNR, SWT, 70 acres. Oak wood.
Glasdrum Wood, head of Loch Creran. NNR, 43 acres.
Glencoe, north-east Argyll. Geological interest with cauldron subsistence; rich woodland flora in alder wood at Carnoch; rare alpines including drooping saxifrage; mountain birds.
Glen Nant, south of Taynuilt. FNR, 104 acres.
Glen Orchy, north of Dalmally. Relict pine woods in Allt Coire Bhiochar and Allt Broighleachan valleys with Highland pine-wood fauna.
Kintyre, peninsular region of south-west Argyll. Sika deer in woodland; varied bird life including eagles and harriers; in winter white-fronted, barnacle and greylag geese on the coastal flats.
Knapdale Forest, south-west Argyll. Roe and sika deer, badgers and otters; black game, woodcock, birds of prey including golden eagle.
Loch Etive. Typical western sea-loch, interesting marine fauna; mergansers, eider, eagles and buzzards at head of loch.
Oban, NM 8630. Rare coastal moths, mountain ringlet in hinterland.
Tighnabruaich Forest, Cowal, south-west Argyll. Photographic hides, information from local Forest Office; roe deer, mountain hares.

Ayrshire

Ballantrae, NX 0882. LNR, SWT, 55 acres. Geological interest with Ordovician fossils; some botanical interest; ternery on shingle.
Dalry. Geological interest with Carboniferous Limestone and Millstone Grit fossils.
Enterkine Wood, north-east of Ayr. NS 4223. LNR, SWT, 12 acres. Mixed semi-natural woodland; badgers; varied bird life.
Glen App, south of Ballantrae. NX 0981. Wooded glen with summer migrants including nightjar.
Hunterston, NS 2154. Waders and some wintering duck.

Banffshire

Banff Coast, from Portknockie to Macduff. Very varied metamorphic rocks of the Dalradian series.
Pennan Head, west of Rosehearty. NJ 8566. Sea-bird colonies.
Troup Head, east of Macduff. NJ 8267. Interesting rock formations; sea-bird colonies.

Berwickshire

Bemersyde Moss, NT 6033. LNR, SWT, 67 acres. Marshland flora and fauna with various aquatic habitats and black-headed gullery.
Cockburnspath, NT 7971. Geological interest—Siccar Point unconformity of great historical interest with Upper Old Red Sandstone resting unconformably on Silurian strata—Carboniferous and Silurian fossils.
Duns Castle, NT 7854. LNR, SWT, 190 acres. Woodland and marshland flora and fauna.
Gordon Moss, NT 6342. LNR, SWT, 101 acres. Peat-bog and birch wood; rich flora and some entomological interest.
Hule Moss, NT 7249. Peat-bog and loch. Noted haunt of pink-footed geese and there are also wintering duck.
St Abb's Head, NT 9169. Sea-bird colonies.

Buteshire

Glen Diomhan, Arran. NR 9346. NNR, 24 acres. Gorge with endemic white-beams.

Clackmannanshire

Cambus, NS 8594. Geological interest with Carboniferous fossils.
Dollar Glen, NS 9699. NTS, 60 acres. Wooded glen in the Ochils with ornithological interest. The Ochil Fault is exposed at the southern end of the glen.
Gartmore Dam, north-east of Alloa, NT 9294. Ornithological interest including wintering greylag geese.

Dumfriesshire

Caerlaverock, Nith Estuary. NNR, 13,514 acres. Haunt of vast numbers of geese in winter, barnacle, greylag and pink-footed; waders and birds of prey.
Caerlaverock Wildfowl Refuge. WT. Wintering geese. Admission charge.
Canonbie, in the valley of the Esk. Carboniferous Limestone and Coal Measure fossils.

Castle and Hightae Lochs, Lochmaben. LNR, DCC, 339 acres. Wintering geese. Interesting aquatic flora and fauna.
Fountainbleau and Ladypark, east of Dumfries. LNR, SWT, 14 acres. Marshy birch wood, botanical and ornithological interest including willow-tits.
Grey Mare's Tail, Moffat, NT 1815. NTS. The area behind, at Dobb's Linn, of considerable geological interest with many Silurian graptolites; the surrounding hills have uncommon alpines; herd of wild goats above the waterfall.
Tynron Juniper Wood, north-west of Dumfries, NX 8093. NNR, 12 acres. An unusual habitat type for southern Scotland.
Wanlockhead, east of Sanquhar, NS 8813. Geological interest with many minerals in the dumps of disused lead-mines.

Dunbartonshire

Ardmore, south-east of Helensburgh, NS 3178. LNR, SWT, 479 acres. Geological interest with a line of unconformity in Old Red Sandstone strata; marine life; various habitats with good variety of plant and animal life.
Loch Lomond. NNR, 624 acres. Geological interest with HBF passing through Inchcailloch; interesting aquatic flora; lochside woodland holds many summer migrants, waders numerous on passage, wintering greylag geese and duck.

East Lothian

Aberlady Bay, NT 4681. LNR, ELCC, 1,439 acres. Geological and ecological interest with varied habitats, rich flora, marine life, excellent birdwatching area.
Barns Ness, east of Dunbar. NT 7277. Geological interest; coastal flora; marine life.
Dunbar, NT 6579. Geological interest with Carboniferous fossils. Estuary of the Tyne good for passage waders, wintering duck and geese.
East Lammermuir Deans, west of Cockburnspath. LNR, SWT, 55 acres. Wooded valleys in the Lammermuir Hills with rich bryophytic flora.
Garleton Hills, near Haddington. Geological interest with Carboniferous volcanic rocks and veins of kidney-ore.
North Berwick, NT 5585. Geological interest with various volcanic rocks along the shore and the volcanic plug of North Berwick Law.
Traprain Law, south-west of Dunbar. NT 5875. Similar hill to North Berwick Law. A variety of minerals in the quarry.

Fifeshire

Bankhead Moss, south-west of St Andrews. LNR, SWT, 9·5 acres. Raised bog.
Cameron Reservoir, south-west of St Andrews. Wintering duck, geese and whooper swans.

Cupar. Geological interest with Old Red Sandstone fossil fish and plant remains.
Eden Estuary, north of St Andrews, NO 4920. Good birdwatching area; waders, terns, wintering duck including large flock of scoter.
Elie Noss, NT 4999. Geological interest with a volcanic neck and garnets known as Elie 'rubies'; bird-migration watch-point.
Fife Ness, NO 6410. Migration watch-point.
Isle of May, NT 6599. NNR, 140 acres. Bird observatory, migration, breeding sea-birds.
Kilconquhar Loch, north of Elie, NO 4901. Botanical and ornithological interest.
Kincardine mud-flats, inner Forth. Waders and wintering duck.
Kinghorn, NT 2787. Geological interest: Carboniferous strata and volcanic rocks.
Largo Bay. Carboniferous Limestone fossils; waders and little gulls.
Morton Lochs, NO 4626. NNR, 59 acres. Interesting aquatic flora and fauna.
Mugdrum Island, Firth of Tay, NO 2319. Roost of pink-footed and greylag geese.
Peppermills Dam, north of Kincardine, NS 9488. Passage migrants.
Tentsmuir Point, NO 4928. NNR, 1,249 acres. Dune flora; terns and many passage migrants.

Inverness-shire

Cairngorms. NNR, 58,822 acres. Superb wildlife area comprising various habitats from natural pine wood to summit plateau. Geological interest with some semi-precious stones such as 'Cairngorm' quartz; botanical interest with rare flowers of pine forest and some uncommon alpines; great ornithological interest with capercaillie, black game, crossbill, crested tit, dotterel, snow-bunting, golden eagle; mammals include red squirrel, wild cat, mountain hare, roe and red deer.
Craigellachie, Aviemore. NNR, 642 acres. Birch woods of botanical and entomological interest with rare moths.
Glen Affric. One of the finest Highland glens with representative wildlife including eagle, buzzard, red and roe deer, wild cat. Remnants of native pine forest and some uncommon mountain plants at head of glen.
Glen Roy. NNR, 2,887 acres. Geological interest in the parallel 'roads'.
Insh Marshes, Spey valley. RSPB reserve. Botanical interest with rich marsh vegetation; passage migration of many species and breeding marsh-birds; otters common.
Loch Garten. RSPB reserve and statutory bird sanctuary. Primarily for ospreys, but other interesting species to be seen.
Monadhliath Mountains. Characteristic Highland fauna including golden eagle and wild cat in remote region less visited than the Cairngorms.

Kincardineshire

Crathes Woods, NO 7396. NTS. Woodland birds.
Girdle Ness, NJ 9705. Migration watch-point.
St Cyrus. NNR, 227 acres. Varied coastal habitats, botanical and entomological interest.

Kinross-shire

Loch Leven, NO 1401. NNR, 3,946 acres. Nesting wildfowl, migration staging post, wintering duck and geese.
Vane Farm, Loch Leven. RSPB reserve and nature centre. Birch woodland flora and fauna, wintering geese.

Kirkcudbrightshire

Cairnsmore of Fleet, NX 5067. Granitic intrusion; black and red grouse; red deer and feral goats.
Carsethorn, NX 9960. Passage waders and large raft of wintering scaup.
Criffel, NX 9661. Granitic intrusion with aureole of various metamorphic rocks.
Dee Valley. Great ornithological interest with large numbers of wintering duck and geese.
Glen Trool National Forest Park. Forest and moorland with rich bird life; red deer and feral goats.
Kirkconnel Flow, NX 9670. NNR, 383 acres. Raised bog, birch and pine woodland.
Kirkcudbright. Silurian fossils.
Knowetops Lochs, Corsock. LNR, SWT, 68 acres. Calcifuge habitats of moorland and loch.
New Abbey. Rich mineral locality at Kinharvie Burn.
New Galloway. Ordovician fossils.
Silver Flowe, NX 4684–781. NNR, 472 acres. Various types of bog.
Southwick, NX 8853. Uncommon minerals.
Threave Wildfowl Refuge, NX 7461. LNR, NTS. Wintering duck and geese. Observation hides.

Lanarkshire

Corehouse, Clyde valley. NS 8842. LNR, SWT, 65 acres. Wooded gorge, and waterfalls.
Hamilton High Parks, NS 7352. Private parkland with famous Cadzow herd of wild white cattle. Application to view should be made to Mr H. R. Brown, Factor, Hamilton and Kinneil Estates Ltd, 18 Auchingramont Road, Hamilton ML3 6JZ.

Hamilton Low Parks, NS 7355. Statutory bird sanctuary. Excellent spot for uncommon passage migrants and rare vagrants but reported to have lessened in interest since construction of the motorway.
Leadhills, NS 8915. Variety of minerals on disused mine dumps.
Muirkirk. Silurian fossils.
Possil Marsh, Glasgow. LNR, SSPWB. Botanical and ornithological interest.
Victoria Park, Glasgow. Fossil forest.

Midlothian

Almondell, NT 0969. Country park of Midlothian County Council abutting river Almond and the Union Canal.
Broadlaw, NT 3453. Granodioritic intrusion and Silurian inlier of radiolarian chert and traptolitic shales.
Dalkeith. Carboniferous fossils including Coal Measure plants.
Edinburgh. Numerous sites of geological interest include the well-known Arthur's Seat, an extinct volcano; Castle Hill crag and tail; Salisbury Crags sill; volcanic rocks at Blackford Pond and Blackford Hill quarries; iron nodules containing fossil fishes on Granton shore; and numerous fossils on Joppa shore. Uncommon flowers of Arthur's Seat include red catchfly and maiden pink. Marshland adjoining Duddingston Loch has been made a SWT reserve. The loch itself is a statutory bird sanctuary with great-crested grebes and various duck. Offshore waters from Leith docks to Musselburgh have vast congregations of wintering duck, especially scaup.
North Middleton Quarry, NT 3559. Disused quarry with numerous Carboniferous Limestone fossils.
Pentland Hills. Silurian and Old Red Sandstone strata with volcanic rocks. Various reservoirs, some with ornithological interest.

Morayshire

Culbin Forest and Sands. Coniferous plantations on old dunes of quite exceptional interest. Rare orchids and dune flora; crossbills, crested tits, capercaillie; red squirrels, roe deer. Wading birds, terns and wintering duck on the coast.
Findhorn Bay. Salt-marsh flora; excellent spot for waders in the early autumn and duck in winter.
Loch Spynie, NJ 2366. Marsh and loch with marshland flora and fauna. Wintering duck and geese. Private estate.
Spey Mouth, NJ 3465. Estuarial area of considerable ecological interest. Terns, passage migrants and wintering duck.

Nairnshire

Nairn. Fossil fish in Old Red Sandstone.
Whiteness Head, NH 8058. Salt-marsh flora; waders and wintering duck.

Peeblesshire

Dawyck Gardens, NT 1836. Private estate with sika deer-park. Occasional opening to the public. Admission charge.
Lamancha, NT 2052. Geological interest, various Ordovician strata and fossils.
Portmore Loch. NT 2650. Ornithological interest throughout the year.

Perthshire

Ben-a-Ghlo, NN 9773. Rare alpines.
Ben Lawers, NN 6544. LNR, NTS. Mountain famous for the wealth of its alpines.
Ben Lui, NN 2627. NNR, 925 acres. Rich mountain flora.
Ben Vrackie, near Pitlochry. Rich mountain flora.
Black Wood of Rannoch, south of Loch Rannoch. Botanical interest, plants include *Pyrola media* and *Listera cordata*; red and roe deer, red squirrels; capercaillie, black grouse, crossbills; rare moths.
Corrycharmaig, Glen Lochay. NN 5236. Disused chromite-mine.
Dunalastair marshes, NN 7158. Marshland flora and fauna.
Flanders Moss, west of Stirling. Bog flora includes Labrador tea-plant; rich bird life includes black grouse, golden plover, woodcock and wintering geese.
Glen Tilt, north Perthshire. Glen Tilt 'marble'.
Keltneyburn, near Fortingall. LNR, SWT, 77·5 acres. Burn and rough grassland of botanical interest.
Loch of Craiglush, NO 0444. LNR, SWT, 88 acres. Ospreys.
Loch of Drumellie, NO 1444. Wintering duck, greylag geese and whooper swans.
Loch of the Lowes, NO 0443. LNR, SWT, 242 acres. Varied habitats with considerable botanical and ornithological interest; rich freshwater life.
Meall nan Tarmachan, NN 5839. NNR, 1,142 acres. Geological and botanical interest.
Old Wood of Meggernie, NN 5545. Relict pine wood; mountain hare, red and roe deer; capercaillie and black grouse. Private estate.
Rannoch Moor. NNR, 3,704 acres. Blanket bog, primarily botanical and entomological but some ornithological interest.
Stormonth Loch, south of Blairgowrie. Wintering duck, greylag geese and whooper swans.
Tyndrum, NN 3230. Disused lead-mines with various minerals on spoil heaps; relict pine wood at Coille Coire Chiule with typical fauna.

Renfrewshire

Castle Semper Loch. LNR, RCC. Wintering duck, greylag geese, whooper swans.
Clyde Muirshiel Regional Park. 30,000 acres in the Renfrewshire Heights. Woodland and moorland. Information Centres at Muirshiel Country Park and Cornalees Bridge.
Paisley. Carboniferous Limestone fossils.

Ross and Cromarty

Achnashellach. FC. Relict pines with birch. A part kept as a deer-sanctuary. Pine martens and wild cats common, a few badgers.
Altt nan Carnan Gorge, NG 8941. NNR, 18 acres. Flora of both calcareous and non-calcareous rocks.
Applecross. Jurassic fossils.
Beauly Firth. Wildfowl at all seasons—especially numerous in autumn.
Beinn Eighe. NNR, 10,507 acres. Mountain and relict pine wood. Eagle, ptarmigan, crossbill; red and roe deer, pine marten and wild cat.
Corrieshalloch, at head of Loch Broom. NNR, 13 acres. NTS and FC. Fine gorge and waterfall; botanical interest.
Coulin, NG 9955. Relict pine woods.
Cromarty Firth. Old Red Sandstone fossils; passage migrants and wintering waders, duck and greylag geese.
Dornoch Firth. Another good birdwatching area for waders and duck—several species of geese on passage.
Eilean na Creige Duibhe, NG 8233. LNR, SWT, 2·8 acres. Sea-loch island with heronry.
Glen Einig, NH 3698. Relict pine wood with capercaillie and black grouse; red, roe and sika deer, pine martens and wild cats.
Inverpolly. NNR, 26,827 acres. Remote district with considerable geological, botanical and ornithological interest.
Munlochy Bay, Moray Firth. Waders, wintering duck, greylag geese.
North Strome Forest, Loch Carron. FC. Several raptor species and pine martens.
Rassal Ashwood, NG 8443. NNR, 202 acres. Limestone supporting a calcareous plant community.
Rhidorroch, NH 2393. Mixed birch and relict pine. Black grouse, red and roe deer, wild cats, pine martens, badgers.
Sgorr Ruadh, NG 9550. One of a number of localities where the dislocation of strata caused by the Moine Thrust can be observed with Torridonian Sandstone on top of Cambrian rocks.
Tarbat Ness, NH 9487. Good birdwatching locality at migration times.

Roxburghshire

Border National Forest Park (part of). Plantations hold woodcock, black grouse and long-eared owl.
Eildon Hills. Range of volcanic hills with the uncommon mineral riebeckite.
Hoselaw Loch and Moss, NT 8132. Marshland flora and fauna.
Yetholm Loch, NT 8028. LNR, SWT, 65 acres. Marshland flora and fauna.

Selkirkshire

Hare and Dunhog Mosses. LNR, SWT, 10 acres. Marshland flora and fauna.
Ettrick Forest. Large area of sparsely populated hilly moorland covering much of the county. Ordovician fossils; moorland birds not plentiful but wild goats occur.
Melrose. Ordovician fossils.

Stirlingshire

Ballagan Glen, NS 5780. LNR, SWT, 12 acres. Gorge in the Campsie Fells; Carboniferous rocks bearing calcareous plant community.
Blairlogie, NS 8397. Geological interest: mineral veins in Old Red Sandstone.
Bridge of Allan, NS 7998. Geological interest: lavas, conglomerate, copper ores and Old Red Sandstone fossils.
Grangemouth, NS 9683. Waders, duck and geese on Forth mud-flats.
Queen Elizabeth National Forest Park. 41,454 acres. Superb scenic area with varied flora and fauna including numerous black grouse and a number of raptor species.
Stirling. Imposing quartz-dolerite sill on which Stirling Castle is built.

Sutherland

Ben Loyal. Geological interest: intrusion of Syenite, an igneous rock type uncommon in Britain.
Brora. Jurassic fossils.
Clo Mor Cliffs. Sea-bird colonies.
Durness. Cambrian fossils.
Faraid Head, NC 3972. Sea-bird colonies.
Gualinn adjoining A 838 south of Durness. NNR. Plant communities of moor and bog.
Inchnadamph. NNR, 3,200 acres. Geological and palaeontological interest in Cambrian strata and cave fossil fauna; rare calcicolous alpines.
Invernaver. NNR, 1,363 acres. Sand-dunes with great botanical interest.
Loch Fleet, east Sutherland. LNR, SWT, 1,270 acres. Botanical interest in dunes and pine wood; waders and wintering duck.

Mound Alderwoods, east Sutherland. NNR, 659 acres. Deltaic fen and alder woodland.
Sandwood Bay, near Cape Wrath. Sand-dunes; interesting minerals.
Stack of Glencoul, NC 2928. Geological interest: Moine Thrust.
Stoer Peninsula. Rich in wildlife. Plants of dune and shingle; varied bird life including greenshank; otters, pine martens, wild cats.
Strathy Bog, NC 7953. NNR, 120 acres. Blanket bog of physiographical and ecological interest.
Strathy Point, NC 8269. Cliff plants.

West Lothian

Dalmeny. Carboniferous fossils.
Linlithgow. Carboniferous fossils; wintering duck on loch.
Pepper Wood, Kirkliston. LNR, SWT, 2·8 acres. Deciduous woodland with rare flowers.

Wigtownshire

Lochinch. Private estate. Wintering duck and greylag geese.
Loch Ryan. Waders, wintering duck and greylag geese.
Maryport, near Mull of Galloway. Rich marine life.
Mochrum and Castle Lochs. Inland cormorant colony and feral greylags.
Mull of Galloway. Sea-bird colonies.
Wigtown Bay. Botanical interest: salt-marsh; marine life; waders, large numbers of wintering duck, greylag geese, whooper swans, pink-footed geese on passage.

Nature trails

Aberdeenshire

Midmar Forest, west of Aberdeen. Two trails in the Tillylair section of the forest. Booklet 5p from FC East Conservancy, 6 Queen's Gate, Aberdeen.

Angus

Arbroath Cliffs, north of town. SWT trail. Geology, sea-birds, cliff flowers, marine life. Booklet 15p (plus postage) from SWT, 8 Dublin Street, Edinburgh.
Camperdown Park, Dundee. Nature centre and trails. Leaflet from Parks and Recreation Dept, 17 City Square, Dundee.

Argyllshire

Caladh Castle, Tighnabruaich. NN 9975. Forest trail 1¼ miles. Guide 7p from FC, 20 Renfrew Street, Glasgow.
Glencoe, NN 1256. Forest walks; Signal Rock trail 1½ miles; Lochan trail 2 miles. Booklet 10p from Head Forester, Glencoe.
Glencoe. NTS guided walks by ranger-naturalist. Bookings on previous day at Information Centre where leaflet can be obtained. Charge for non-members.
Inverinan and Inverliever, Loch Awe. NN 0018. One nature trail and four forest walks. 1¼, 2, 2¼, 4½, 5 miles. Booklet 10p from Chief Forester, Inverliever Forest, by Taynuilt, Argyll.
Knapdale, near Crinan Basin. Wildlife exhibition hut. Forest walks. Leaflet from Forester, Knapdale Forest.
Strontian Glen. NC trail. Booklet 3p from Strontian Village Centre or NC.

Ayrshire

Culzean Country Park, Maidens. March to October. Trails ¾–1½ miles. Leaflet at Information Centre, Culzean Castle.
Enterkine Wood. ½ mile. Guide and permits from Mr J. Lorrain-Smith, 22 Auchenbeg Crescent, Ayr.
Rozelle, Ayr. ½ mile. Booklet 6p from Director of Parks, 30 Miller Road, Ayr.

Buteshire

Loch Fad, Bute. 4½ miles. Guide 5p from Rothesay Museum.
Loch Ascog, Bute. 1¼ hours. Guide 5p from Rothesay Museum.
Ettrick Bay to Kilmichael, Bute. 5 or 10 miles. Guide 5p from Rothesay Museum.
Kingarth, Bute. 2–2½ hours. Marsh and seashore. Guide 5p from Rothesay Museum.
South End of Bute. 3 hours. Guide 10p from Rothesay Museum.
Motorists' Trail, Bute. 30–35 miles. Guide 10p from Rothesay Museum.
Bull Loch, Bute. 2–3 hours. Guide 10p from Rothesay Museum.

Dumfriesshire

Ae Forest Walks. Start from Forest Office near Ae village. 3¼ and 3½ miles. Wildlife includes roe and fallow deer. Booklet 6p from Conservator, FC, Greystone Park, Moffat Road, Dumfries.
Grey Mare's Tail, near Moffat. NTS guided walks by ranger-naturalist. Leaflet from Countryside Adviser, NTS, 5 Charlotte Square, Edinburgh.

Dunbartonshire

Ardmore, near Helensburgh. SWT coastal trail, marine life and geological

interest. Booklet 15p from Clyde area branch secretary, Mr J. M. Findlay, 1 Westbank Quadrant, Glasgow W2.
Inchcailloch, island on Loch Lomond. NC trail. Leaflet 3p from warden Mr A. J. MacFarlane, The Boatyard, Balmaha.
Loch Lomond Park. ¾–1 mile. Start from Nature Trail Information Hut. Guide from Supt's House, Balloch.

East Lothian

Barns Ness, near Dunbar. Geological trail, ½ mile. Guide 10p from ELCC, County Buildings, Haddington.
Dunbar. Start at Dunbar rocks, west of harbour. 1¾ miles. Information from District Clerk, Town House, Dunbar.
Pressmennan, East Linton, NT 6172. Forest trail 2 miles. Leaflets from FC district office, Glentress, Peebles.
Yellowcraig, near North Berwick. ¾ mile. Guide 5p from ELCC, County Buildings, Haddington.

Fifeshire

Dunnikier Park, Kirkcaldy. 1 mile. Guide 5p on site or Parks Dept, Town House, Kirkcaldy.
Ravenscraig Park, Kirkcaldy. 1 mile. Guide 2½p on site or Parks Dept, Town House, Kirkcaldy.
Tayport. Two forest nature trails constructed by Tayport Preservation and Development Society.

Inverness-shire

Achlean, Glen Feshie. NC trail includes deer observation tower. Guide from Achlean Croft.
Craigellachie, Aviemore. NC trail. 1 mile. Guide from Aviemore Centre.
Craig Phadrig, Inverness. Forest trail. Leaflet 5p from FC, 60 Church Street, Inverness.
Culloden, east of Inverness. Forest trail. Starts at Smithton. Guide from FC as above.
Farigaig, 15 miles north-east of Fort Augustus. Forest trail. Leaflet 5p from Forest Office.
Glenmore, near Coylum Bridge. Eight trails. Leaflet 5p and map 10p from camp-site.
Inchnacardoch, Fort Augustus. Two forest trails. Leaflet 5p from Forest Office.
Landmark, near Carrbridge. Trail in grounds of Landmark Visitors' Centre.
Loch an Eilean, Rothiemurchus Forest. NC trail 2½ miles. Leaflet on site.

Reelig Glen, 8 miles west of Inverness. Forest walk with geological and botanical interest.

Kincardineshire

Banchory (5 miles west of). Forest trail 3¼ miles. Guide 5p from FC, 6 Queen's Gate, Aberdeen.
Crathes Castle, Banchory. NTS trail. 1¾ miles. Guide on site.
St Cyrus. NC trail. 2½ miles. Guide 3p from NC, 12 Hope Terrace, Edinburgh.

Kinross-shire

Vane Farm, Loch Leven. RSPB trail. 1½ miles. All year but not Mon or Fri. 0930–1730. Guide from Warden on site.

Kirkcudbrightshire

Fleet, NX 6056. Three forest walks, 1, 1½ and 2½ miles. Start from Murray Forest Centre. Booklet 5p from Forest Office, Gatehouse of Fleet or FC, Greystone Park, Moffat Road, Dumfries.
Glen Trool Forest Park. Several forest trails, Loch Trool, Kirroughtree, Stroan Bridge.

Lanarkshire

Corehouse, 2 miles from Lanark. SWT trail, 1–2 miles. Woodland and riverside trail. Guide 30p and permit from Secretary, Corehouse Nature Reserve, 18 Ladyacre Road, Lanark.
East Kilbride, NS 6552. 1 mile. Details from Director of Parks, Civic Centre, East Kilbride.
Neilsland Park, Hamilton. Woodland wild flowers. Booklet 5p from Parks Dept, Burgh of Hamilton.
Orbiston Glen, Bellshill, NS 7459. 1¾ miles along banks of South Calder Water. Guide from Parks Superintendent, Council Chambers, Bellshill.

Midlothian

Almondell. Starts at South Lodge, East Calder. Riverside and canal walk. Booklet from Publications Officer, Midlothian County Council, George IV Bridge, Edinburgh.
Hermitage and Blackford trail, Edinburgh. 1½–2 miles. Starts from Hermitage of Braid public park. Mainly for children.
Pentlands. NTS guided walks by ranger-naturalist. Further information and leaflet from NTS, 5 Charlotte Square, Edinburgh.

Morayshire

Monaughty Forest, Torrieston, NJ 1659. Forest walks 1–2 miles. Further information from FC, 6 Queen's Gate, Aberdeen.

Peeblesshire

Glentress Forest, 2 miles east of Peebles. 4¾ miles. Booklet 5p on site or FC, Greystone Park, Moffat Road, Dumfries.

Perthshire

Atholl Way. Motorists' nature trail from Dunkeld to Blair Atholl. Booklet 10p from Blair Castle and NTS Information Centres at Dunkeld and Killiecrankie.
Ben Lawers. Trail on lower slopes. Starts from NTS Information Centre. 1½–2 miles. Leaflet on site. Also guided walks by ranger-naturalists from the Centre where leaflet can be obtained. Mountain trail 7 miles. Guide 5p from Centre.
Birks of Aberfeldy. SWT trail on banks of Urlar Burn. Booklet 5p from Town Council, Municipal Chambers, Aberfeldy.
Craigvinean, near Dunkeld. Forest walks, 1, 3¼ and 8 miles. Guide 10p from FC, 6 Queen's Gate, Aberdeen.
Crieff, west of town. SWT trail along banks of river Turret. 1½ miles. Leaflet 5p from Crieff Hydro.
Drummond Hill, Kenmore. FC walks 3 miles. Further information from Chief Forester, Dalerb, Loch Tay, Aberfeldy.
Dunkeld. Various guided walks by ranger-naturalist between Dunkeld and Blair Atholl. Charge for non-members of NTS. Further information and leaflet from NTS Visitors' Centre, Killiecrankie.
Faskally, Pitlochry. 1½ miles. Guide 10p from Forest Information Centre on site.
Kindrogan, Enochdhu, Strathardle, NO 0563. Field Study Centre trail. 1 mile. Guide 8p from Warden, Kindrogan Field Centre, Enochdhu, Blairgowrie. Hill trail, 3 miles also starting from Centre. Guide 8p.
Kinnoull Hill, Perth, NO 1323. 1 mile. Starts from Hatton Road Lodge. Guide 5p including postage from Director of Parks, Marshall Place, Perth.
Linn of Tummel, Pitlochry. 2¼ miles. NTS trail. Booklet 10p from NTS Information Centres at Dunkeld and Killiecrankie.
Loch Ard, 2 miles south of Aberfoyle. Forest walks in Loch Ard section of Queen Elizabeth Forest Park. 1½, 3, 6, 7½ and 8 miles. Start from Cobleland caravan site. Trail leaflet 5p from David Marshall Lodge, Aberfoyle.
The Hermitage, Dunkeld. NTS trail by river Braan. 1½ miles. Guide 5p from NTS Visitors' Centre, Dunkeld.
Tummel Forest, Pitlochry. Forest walks. Booklet 10p from FC Tummel Forest Office.

Renfrewshire

Cornalees Bridge, near Inverkip. Trail through a SSSI designated glen. Leaflet from Director of Planning and Engineering, County Buildings, Cotton Street, Paisley.
Muirshiel Country Park, north-west of Lochwinnoch. Adult and junior trails. Start from Information Centre. Guide from Centre or from Director of Planning as above.

Ross and Cromarty

Beinn Eighe, Kinlochewe. Forest trail in Glas Leitire wood and mountain trail on Beinn Eighe. Leaflets from Warden at Anacaun Field Station, Nature Conservancy, Kinlochewe.
Inverpolly. Motorists' trail. 50-mile drive starting and finishing at Knockan car park. Booklet 10p from Information Centre at car park.
Kintail and Balmacara. NTS guided walks by ranger-naturalist from Morrich and Balmacara Information Centres where leaflet can be obtained. Charge for non-members.
Knockan Cliff. NC geological trail, 1½ miles. Leaflet from Information Centre on site.
Slattadale Forest. FC forest walks 1 mile and 5 miles. Guide from FC, 60 Church Street, Inverness.
Torrachilty, Strathpeffer. Forest walk. Leaflet 5p from Forestry Office.
Torridon. NTS guided walks by ranger-naturalist. Bookings at least 24 hours before at Countryside Centre, Torridon where leaflet can be obtained. Charge for non-members.

Stirlingshire

Abbey Craig, 1½ miles north of Stirling. Guide 5p from Tourist Centre, Stirling

Sutherland

Kyle of Sutherland, Bonar Bridge. Four walks. Booklet 5p on site.

West Lothian

Hopetoun House, Bo'ness. Nature trail through grounds of house. Booklet 15p on site or from SWT, 8 Dublin Street, Edinburgh.

Wigtownshire

Kilsture, Sorbie, NX 4448. Forest walk, 1 mile. Guide 7½p from Forest Office or FC, Greystone Park, Moffat Road, Dumfries.

Organisations, societies, field study and other centres, research laboratories and wildlife park

Aberdeen and North of Scotland Zoological Society, c/o Natural History Department, Marischal College, Aberdeen.

Aberdeen Natural History and Antiquarian Society.

Andersonian Naturalists of Glasgow, c/o Mr J. D. Scobie, 7 Balmuildy Road, Glasgow.

Banchory Research Station, Nature Conservancy.

Ben Lawers NTS Visitors' Centre.

Berwickshire Naturalists' Club.

Botanical Society of Edinburgh, Royal Botanic Garden, Inverleith Row, Edinburgh.

Buteshire Natural History Society, Miss D. N. Marshall, The Museum, Stuart Street, Rothesay, Bute.

Committee for the Study of the Scottish Flora, Mr R. Mackecknie, 9 Skirving Street, Glasgow S1.

Dumfriesshire and Galloway Natural History and Antiquarian Society, Mrs H. R. McAdam, Maryfield, Bankend Road, Dumfries.

Dundee Naturalists' Society, Mrs H. M. Brown, 1,169 Turnberry Avenue, Dundee.

Dunfermline Naturalists' Society, Mrs E. M. Stewart, 99 Halbeath Road, Dunfermline, Fife.

East Lothian Antiquarian and Field Naturalists' Society.

Edinburgh Natural History Society.

Fort William Natural History Society.

Garth Field Study Centre, Glen Lyon, Perthshire.

Hamilton Natural History Society, Mrs A. Wallace, Wingfield, 28 South Park Road, Hamilton, Lanarkshire.

Highland Wildlife Park, Kincraig, Inverness-shire. Collection of Scotland's wildlife, past and present. Daily 1000–1800. Admission charge.

Inverness Botany Group.

Inverness Birdwatching Group.

Isle of May Bird Observatory. Managed by a Joint Committee. Further information from the Regional Officer for North and Central Scotland, The Nature Conservancy, 12 Hope Terrace, Edinburgh.

Kindrogan Field Study Centre, Enochdhu, Strathardle, by Blairgowrie, Perthshire.

Kirkcaldy Naturalists' Society, Mr J. B. Bease, 55 King Street, Kirkcaldy, Fife.

Landmark Visitors' Centre, Carrbridge, Inverness-shire. Original exhibition portraying history of Scottish landscape and wildlife. Admission charge.

Mid-Argyll Natural History and Antiquarian Society, Mr F. Bruce, Auchindarroch Hotel, Ardrishaig, Argyll.

Midlothian Ornithologists' Club.

Millport Marine Biological Station, Great Cumbrae. Robertson Museum and Aquarium open to the public for small admission charge.

Nature Conservancy, Scottish HQ, 12 Hope Terrace, Edinburgh.

Perthshire Society of Natural Science, Mr A. W. Robson, Towerview, Dunning.

Red Deer Commission, Elm Park, Island Bank Road, Inverness.

Renfrewshire Natural History Society, Mr A. Clarkson, 12 Dargavel Avenue, Bishopton.

Royal Society for the Protection of Birds, Scottish Office, 17 Regent Terrace, Edinburgh.

Scottish Department of Agriculture and Fisheries Freshwater Laboratory, Faskally, Pitlochry.

Scottish Department of Agriculture and Fisheries Marine Laboratory, Victoria Road, Torry, Aberdeen.

Scottish Field Studies Association, Estate Office, Blair Drummond, by Stirling.

Scottish Marine Biological Association Research Laboratory, Dunstaffnage, Oban, Argyllshire.

Scottish Ornithologists Club, 21 Regent Terrace, Edinburgh.

Scottish Wildlife Trust, 8 Dublin Street, Edinburgh.

Torridon NTS Countryside Centre. Wildlife exhibits, slide show, booklets for sale.

Museums

Aberdeen University Natural History Museum, Zoology Dept, Tillydrome Avenue. By appointment only. Tel 40241.

Airdrie Public Museum, Lanarkshire. Weekdays 1000–2100. Collection of David Stephen's annotated wildlife photographs.

Annan Museum, Moat House, Annan. Weekdays 0900–1700.

Anstruther, Fifeshire. The Scottish Fisheries Museum. Has marine aquarium. June–Sept weekdays 1000–1230, 1400–1800, Sun 1400–1700. Oct–May daily, except Tues, 1430–1630. Admission charge.

Brechin Museum, Mechanics Institute. Tues and Wed 1400–1700, Sat 1030–1230.

Campbeltown Museum. Includes geology and natural history of Kintyre. Weekdays 1015–1800 (1015–1300 Wed). Open until 2100 during summer.

Hugh Miller's Cottage, Cromarty Firth. Geological specimens. April–Oct weekdays 1000–1200, 1300–1700 (Sun 1400–1700 from June). Admission charge, children free if accompanied.

Dumfries Burgh Museum. Good mineral collection. Weekdays except Tues 1000–1300, 1400–1700 and April-Sept Sun 1400–1700.

Dundee City Museum and Art Galleries. Includes natural history, botanical and geological material. Weekdays 1000–1730.

Dundee Broughty Castle Museum, Broughty Ferry. Includes natural history of the Tay. Weekdays except Fri 1000–1300, 1400–1700, Sun 1400–1700.

Dunfermline Museum. Includes regional natural history. Mon-Fri except Tues 1100–1300, 1400–1930, Sat 1100–1300, 1400–1700, Sun 1400–1700.

Royal Scottish Museum, Chambers Street, Edinburgh. Superb wildlife collections, some in natural habitat settings. Weekdays 1000–1700, Sun 1400–1700.

Forfar, the Meffan Institute Museum. Includes geological and natural history material. Weekdays 1000–1200, 1400–1600, Sat 1000–1200.

Forres, the Falconer Museum. Fossils from the Culbin Sands. Summer 1000–1700, winter 1000–1500. Admission charge but Sat free.

West Highland Museum, Fort William. Weekdays Sept-May 0930–1700. June-Aug 0930–2100. Admission charge. Small wildlife section but re-organisation envisaged.

Galloway Deer Museum, Clatteringshaws Loch, 6 miles west of New Galloway. Not restricted to deer: includes goats, birds of prey, botany and geology.

Glasgow Art Gallery and Museum, Kelvingrove. Animals of Scotland, British Bird Gallery, Geology Gallery, Zoological Gallery. Weekdays 1000–1700, Sun 1400–1700.

Camphill Museum, Queen's Park, Glasgow. Includes natural history material. Weekdays 1100–1700, Sun 1400–1700.

The Hunterian Museum, Glasgow University. Includes geological collections. Mon-Fri 0900–1700, Sat 0900–1200.

Glenesk Museum, Angus. Daily June-Sept 1400–1800, winter 1400 to dusk. Admission charge.

Greenock, the McLean Museum, 9 Union Street, West End. Includes natural history and geology. Mon-Sat 1000–1700.

Inverurie Museum, Public Library Building. Small geological and natural history collection. Weekdays 0930–1200, 1500–1700.

Inverness Museum and Art Gallery, Castle Wynd. A small wildlife section. Weekdays 0900–1700 (Weds 0900–1300, 1415–1700).

Kilmarnock, the Dick Institute Museum, Elmbank Avenue. Includes geological and ornithological collections. Oct-April 1000–1700, May-Sept 1000–2000 (Wed and Sat 1000–1700).

Kirkcaldy Museum and Art Gallery, War Memorial Grounds. Includes local geological collection. Weekdays 1100–1700, Sun 1400–1700.

The Stewartry Museum, St Mary Street, Kirkcudbright. Regional museum. Daily 1000–1700. Admission charge.

Robertson Museum and Aquarium, Millport, Great Cumbrae. Weekdays 0930–1230, 1400–1700 (closed Sat in winter). Admission charge.

Montrose Museum, Panmure Place. Regional collection including natural history and geology. Daily 1000–1300, 1430–1630. Admission charge.

North Berwick, the Burgh Museum. Includes natural history. May-Sept. Daily 1400–1700.

Paisley Museum and Art Galleries. Includes Renfrewshire geology and natural history. Weekdays 1000–1700 (1000–2000 on Tues and 1000–1800 Sat).

Peebles Chambers Institute, High Street. Flora, fauna, geological specimens and other natural history exhibits. Weekdays 1000–1900 (Wed 1000–1200) (Sat 1000–1700).

Perth Art Gallery and Museum, George Street. Includes regional natural history. Mon-Fri 1000–1300, 1400–1700 (also Wed and Fri 1800–2000). Sat 1000–1300, 1400–1700, Sun 1400–1600.

Rothesay, Buteshire Natural History Society Museum. April-Oct weekdays 1030–1230, 1430–1630 (also Sun June-Sept 1430–1630). Nov-March weekdays 1430–1630. Admission charge.

Rozelle Mansion House, Ayr. Wildlife Exhibition staged by Royal Burgh of Ayr in conjunction with SWT.

Saltcoats, North Ayrshire Museum. Some local fossils. Summer weekdays 1000–2100, Sun 1400–1700. Winter Sat 1100–1700. Admission charge.

Stranraer, Wigtown County Museum. Natural history section is mainly birds; small collection of local fossils and shells. Some extension of wildlife exhibits planned. Library hours.

Bibliography

Anthony, J. 'Contribution to the Flora of Sutherland, Bettyhill', *Trans Bot Soc Edin*, 38 (1959), 7–15

Alexander, W. B. 'The Natural History of the Firth of Tay', *Trans and Proc Perth Soc Nat Sc*, 9 (1932), 35–42

Barnes, H. and Stone, R. L. 'New Record for *Elminius modestus* Darwin in Western Scotland', *Crustaceana*, 23, pt 3 (1972), 309–10

Beirne, D. P. *Origin and History of the British Fauna* (1952)

Baxter, E. V. and Rintoul, L. J. *The Birds of Scotland* (Edinburgh, 1953)

Brookes, Brian S. 'A Bryophyte Flora of Handa Island', *Trans Bot Soc Edin*, 37 (1957), 114–22

Brown, Ian R. and Tuley Graham. 'A Study of a Population of Birches in Glen Gairn', *Trans Bot Soc Edin*, 41 (1971), 231–45

Brown, P. W. 'Feral Mink in Scotland', *Scottish Agriculture*, 49, no 4 (1970), 163–6

Buchanan, D. 'River Pollution in Lanarkshire', *Hamilton Nat Hist Soc 2nd Report* (1972), 9–11

Burnett, J. H. (ed). *The Vegetation of Scotland* (Edinburgh, 1964)

Campbell, Bruce. *Bird-watching in the West Highlands*, RSPB Occ Publications no 13 (1950)

Chapman, V. J. *Salt-Marshes and Salt-Deserts of the World* (1960)

Charles, W. N. 'The Effect of a Vole Plague in the Carron Valley, Stirlingshire', *Scot Forestry*, 10 (1956), 201–4

Cowper, C. N. L. 'Breeding Distribution of Grey Wagtails,

Dippers and Common Sandpipers on the Midlothian Esk', *Scottish Birds*, 7, no 6 (1973), 302–6

Crooke, Myles and Kirkland, R. C. 'The Gale of 1953: An Appraisal of Its Influence on Forest Pest Populations in Pine Areas', *Scot Forestry*, 10 (1956)

Crundwell, A. C. 'The Recent Discovery in Perthshire of Two Bryophytes New to the British Isles', *Trans and Proc Perth Soc Nat Sc*, 11 (1966), 30

Darling, F. Fraser and Boyd, J. Morton. *The Highlands and Islands* (1964)

Delany, M. J. 'Ecological Distribution of Small Mammals in North-West Scotland', *Proc Zoo Soc Lond*, 137 (1961), 107–26

——. 'The Mammal Fauna of the Clyde Area', *Glasgow Naturalist*, 18, pt 1 (1958), 15–27

——. 'The Small Mammals of a Dunbartonshire Oakwood', *Glasgow Naturalist*, 17, pt 6 (1956), 272–8

Dept of Agriculture and Fisheries for Scotland, *Fisheries of Scotland—Report for 1971*, Cmnd 4,957 (1972)

Eggeling, W. J. *Isle of May* (1960)

Egglishaw, H. J. and Morgan, N. C. 'A Survey of the Bottom Fauna of Streams in the Scottish Highlands II: The Relationship of the Fauna to the Chemical and Geological Conditions', *Hydrobiologia*, 26 (1965), 173–83

Fletcher, W. W. and Martin, D. J. 'Flora of Great Cumbrae Island', *Trans Bot Soc Edin*, 39 (1960), 46–61

Forman, Bruce. 'Spiders of Aberdeenshire', *Scot Nat*, 63, no 3 (1951), 137–55

Gibb, Dorothy C. 'Survey of the Commoner Fucoid Algae', *Jnl Ecol*, 38 (1950)

Gibson, J. A. 'The Mammals of the Island of Arran', *Trans Buteshire Nat Hist Soc*, 17 (1969), 59–68

——. 'The Reptiles and Amphibians of the Island of Arran', op cit, 69–72

——. 'The Mammals, Reptiles and Amphibians of Ailsa Craig', op cit, 73–6

——. 'The Mammals of the Island of Bute', *Trans Buteshire Nat Hist Soc*, 18 (1970), 5–20
——. 'The Reptiles and Amphibians of the Island of Bute', op cit, 31–2
——. 'The Reptiles and Amphibians of Sanda, Sheep Island and Glunimore', op cit, 33–4
——. 'The Mammals of Sanda, Sheep Island and Glunimore', op cit, 48–50
Gimingham, C. H. and Robertson, C. T. 'Contributions to the Maritime Ecology of St Cyrus, Kincardineshire II: Sand Dunes', *Trans Bot Soc Edin*, 35 (1951), 370
Gordon, Seton. *Highland Summer* (1971)
Hudson, Robert. 'The Spread of the Collared Dove in Britain and Ireland', *Brit Birds*, 58, no 4 (1965), 105–39
Hunter, W. Russell and Hunter, M. Russell. 'Mollusca of Scottish Mountains', *Jnl Conch*, 24, no 3 (1956), 80
Jenkins, D. 'The Present Status of the Wild Cat in Scotland', *Scot Nat*, 70 (1962), 126–38
Knight, J. E. and Sutton, F. R. 'Lepidoptera of the Beinn Eighe Nature Reserve', *Ent Gaz*, 17 (1966), 125–8
Lewis, J. R. 'Intertidal Communities of the Northern and Western Coasts of Scotland', *Trans Roy Soc Edin*, 63 (1956), 185–220. *The Ecology of Rocky Shores* (1964)
Lockie, J. D. 'The Breeding Habits and Food of Short-Eared Owls after a Vole Plague', *Bird Study*, 2 (1955), 53–69
——. 'The Breeding Density of the Golden Eagle and Fox in Relation to Food Supply in Wester Ross, Scotland', *Scot Nat*, 71 (1964), 67–77
McLellan, R. *The Isle of Arran* (Newton Abbot, 1970)
MacNally, Lea. *Highland Deer Forest* (1970)
——. *Highland Year* (1972)
McVean, D. N. and Ratcliffe, D. A. *Plant Communities of the Scottish Highlands* (1962)
Maitland, P. S. 'Aquatic Fauna of the Isle of May', *Trans D'fries and G'way Nat Hist and Ant Soc*, 44 (1967), 16–28

——. 'Fish in South-West Scotland', *Trans D'fries and G'way Nat Hist and Ant Soc*, 47 (1970), 49–62
——. *Studies on Loch Lomond—The Fauna of the River Endrick* (1966)
——. 'Echo Sounding Observations on the Lochmaben Vendace', *Trans D'fries and G'way Nat Hist and Ant Soc*, 44 (1967), 29–46
Mitchell, G. H., Walton, E. K. and Grant, D. (ed). *Edinburgh Geology* (1960)
Morgan, N. C. and Egglishaw, H. J. 'A Survey of the Bottom Fauna of Streams in the Scottish Highlands I: Composition of the Fauna', *Hydrobiologia*, 25 (1965), 181–211
Murray, W. H. *Highland Landscape* (Edinburgh, 1962)
Nethersole-Thompson, D. *The Greenshank* (1951)
——. *The Snow Bunting* (Edinburgh, 1967)
Ovington, J. D. 'Afforestation of the Culbin Sands', *Jnl Ecol*, 38 (1950), 303
Perkins, E. J. 'Marine Fauna and Flora of the Solway Firth Area, Pt I', *Trans D'fries and G'way Nat Hist and Ant Soc*, 45 (1968), 15–43
'Pt II', *Trans D'fries and G'way Nat Hist and Ant Soc*, 46 (1969), 1–26
'Pt III', *Trans D'fries and G'way Nat Hist and Ant Soc*, 48 (1971), 12–68
Perry, R. *In the High Grampians* (1948)
Pyefinch, K. A. *Trout in Scotland* (Edinburgh, 1960)
Ratcliffe, D. A. 'The Peregrine Population of Great Britain in 1971', *Bird Study*, 19, no 3 (1972), 117–56
Raven, John and Walters, Max. *Mountain Flowers* (1956)
Richmond, K. *A Regional Guide to the Birds of Scotland* (1968)
Richter, R. 'The Aquatic Coleoptera of the County of Elgin', *Scot Nat*, 63 (1951), 101–21
Ritchie, J. *Influence of Man on Animal Life in Scotland* (Cambridge, 1919)
Ritchie, J. C. 'Primula Scotica—Account for Biological Flora of the British Isles', *Jnl Ecol*, 42 (1954), 623–8

Robertson, E. T. 'Contribution to the Maritime Ecology of St Cyrus, Kincardineshire. 1 The Cliffs', *Trans Bot Soc Edin*, 35 (1951), 370–87

Roy, A. B. 'The Spiders of the Black Wood of Rannoch', *Scot Nat*, 67 (1955), 19–22

——. 'The Spiders of the Cairngorm Region', *Scot Nat*, 70 (1962), 96–101

Shorten, M. 'Squirrels in England, Wales and Scotland', *Jnl A Ecol*, 26 (1957), 287–94

Simms, E. *Woodland Birds* (1971)

Slack, H. D. *Studies on Loch Lomond* (1957)

Southern, H. N. and Reeve, E. C. R. 'Quantitative Studies in the Geographical Variation of Birds—The Common Guillemot (*Uria aalge*)', *Proc Zool Soc Ser A*, 3 (1941), 255–76

Statistical Account of Scotland (3rd). Vols on Aberdeenshire (1960), Argyllshire (1961), Ayrshire (1951), Banffshire (1961), Dumfriesshire (1962), Dunbartonshire (1959), East Lothian (1953), Edinburgh (1966), Fifeshire (1952), Kirkcudbrightshire and Wigtownshire (1965), Lanarkshire (1960), Morayshire and Nairnshire (1965), Peeblesshire and Selkirkshire (1964)

Steers, J. A. 'The Coastline of Scotland', *Geographical Journal*, 118, 150

Steven, H. M. and Carlisle, A. *The Native Pinewoods of Scotland* (Edinburgh, 1959)

Tegner, H. *A Naturalist on Speyside* (1971)

Tittensor, R. M. 'Historical Ecological Changes on Loch Lomond', *Trans Bot Soc Edin*, 41, pt 2

Vevers, H. G. 'Land Vegetation of Ailsa Craig', *Jnl Ecol*, 24 (1936), 424

Watson, A. and Hewson, R. *Mountain Hares* (1963)

Watt, A. S. 'Preliminary Observations on Scottish Beechwoods', *Jnl Ecol*, 19 (1931), 137–57

Watt, A. S. and Jones, E. W. 'Ecology of the Cairngorms', *Jnl Ecol*, 36 (1948), 283–304

Weir, Tom. *Scottish Lochs*, 2 vols (1972)
Whitehead, G. K. *Wild Goats of Great Britain and Ireland* (Newton Abbot, 1972)
Yapp, W. B. 'Oaks in Scotland', *Scot Nat*, 70 (1961,) 2–6
——*Birds and Woods* (1962)

Acknowledgements

The staffs of the Council for Nature, the Scottish Wildlife Trust and the Scottish Marine Biological Association have courteously responded to my requests for information. Mr J. F. McSorley and Mr D. F. C. Forbes of the South Scotland Conservancy of the Forestry Commission kindly answered queries relating to their region and Mr T. G. Haigh of the Department of Agriculture and Fisheries in Edinburgh went to some trouble to supply me with an account of the present distribution of mink.

Mrs A. Wallace, Secretary of Hamilton Natural History Society, supplied me with helpful information and the curators of a number of museums gave me details of their wildlife exhibits.

I am indebted to Mr G. Darby, Editorial Manager of the *Sunday Times* for permission to reproduce the distribution map of the mountain hare and to Mr Lea MacNally for providing four of his fine photographs.

The staff of the National Library of Scotland in Edinburgh assisted my research and I was glad to avail myself of the facilities of the Scottish Ornithologists Club library.

My daughter assisted in the tracing of the maps; my wife in addition has helped with the checking and revision and has given me valuable advice on the numerous problems which inevitably arise in the writing of a book. To them both I give my grateful thanks.

Index

Italic figures indicate illustrations